Inactivity: Physiological Effects

Inactivity: Physiological Effects

Edited by

Harold Sandler

Cardiovascular Research Office
Biomedical Research Division
National Aeronautics and Space Administration
Ames Research Center
Moffett Field, California
and
Stanford University School of Medicine
Stanford, California

Joan Vernikos

Cardiovascular Research Office
Biomedical Research Division
National Aeronautics and Space Administration
Ames Research Center
Moffett Field, California
and
Department of Pharmacology
Wright State University School of Medicine
Dayton, Ohio

1986

ACADEMIC PRESS, INC.
Harcourt Brace Jovanovich, Publishers
Orlando   San Diego   New York   Austin
Boston   London   Sydney   Tokyo   Toronto

ACADEMIC PRESS, INC.
Orlando, Florida 32887

United Kingdom Edition published by
ACADEMIC PRESS INC. (LONDON) LTD.
24–28 Oval Road, London NW1 7DX

Library of Congress Cataloging in Publication Data

Inactivity : physiological effects.

    Includes bibliographies and index.
    1. Hypokinesia—Physiological effect.   2. Stress
(Physiology)   I. Sandler, Harold, Date
II. Vernikos, Joan.   [DNLM: 1. Exertion.
2. Immobilization.   WE 103 I35]
QP310.5.I53   1986   612   86-10937
ISBN 0—12—618510—7 (alk. paper)

PRINTED IN THE UNITED STATES OF AMERICA

86 87 88 89   9 8 7 6 5 4 3 2 1

Contents

6 Psychosocial and Chronophysiological Effects
of Inactivity and Immobilization 123

Charles M. Winget and Charles W. DeRoshia

7 Exercise Responses after Inactivity 149

Victor A. Convertino

8 Conclusions 193

Harold Sandler and Joan Vernikos

Foreword

The Surgeon General of the United States once said, "Smoking is bad for your health," and that message now appears on every cigarette package in the United States.

He could also have said, "Lying in bed is bad for your health." Perhaps this label should be attached to all beds. In fact, we might say that physiologically the most dangerous activity to indulge in is inactivity.

During the Second World War, Sir David Cuthbertson drew attention to the fact that individuals lying in bed, with fractures of one of the major bones of the leg, excreted more calcium and nitrogen than normal, ambulant people. Then he studied some normal, intact people lying in bed for the same period of time and found a similar increase in the excretion of calcium and nitrogen. So this phenomenon did not result from the effects of the fracture as he had thought, but from the effects of lying immobilized in bed.

People who spent a long period in bed because of a fracture, surgery, or some infectious disease always found themselves weak and dizzy when they first attempted to sit up or stand. This was attibuted to weakness brought on by the injury or infection. In fact, normal people lying in bed for the same time have the same symptoms when they try to sit up or stand. So what, in fact, happened was that lying horizontally in bed was causing a deconditioning of the cardiovascular system that created an orthostatic hypotension when they reassumed the erect posture.

Then, when humans began to be exposed to the weightless state in space, it was found that there was increased loss of calcium and nitrogen from their bodies and a cardiovascular deconditioning that became apparent when they returned to a 1 G environment. In many ways, weightlessness resembles bed rest, and thus bed rest experiments became important as a form of ground study of the physiological effects of weightlessness. The study of bed rest also became important as a tool with clinical applications in studying conditions such as osteoporosis and the physiology of the cardiovascular system. One of the methods used to study the physiology of weightlessness in space was to subject humans to sustained bed rest and then spin them in a centrifuge at high G forces. In this way, it was possible to simulate an astronaut's spending some time in space and then being subjected to the higher G forces of reentry.

The two editors of this book have played an integral part in the discoveries just outlined and are uniquely qualified for the job they have taken on. In fact, it would be difficult to find two better qualified individuals for this purpose.

Harold Sandler received his M.D. in 1955 from the University of Cincinnati. After his internship and appointment as a research fellow in medicine, he was introduced to an area he was to make his own for the next 25 years. This was a military service appointment with the U.S. Naval Air Development Center's Aviation Medical Acceleration Laboratory in Johnsville, Pennsylvania. Here Dr. Sandler was brought in contact with the effects of G forces on the cardiovascular system, a system about which he had already published a number of articles. He next held two appointments as assistant professor in medicine and in 1970 joined the National Aeronautics and Space Administration Ames Research Center at Moffett Field, California, as research scientist and the next year was appointed assistant clinical professor of medicine at Stanford University. He has remained at NASA-Ames ever since. From 1972 to 1985, he served as chief of the Bio-medical Research Division. Currently, he is chief of the Cardiovascular Research Office at NASA-Ames; associate clinical professor, Wright State University School of Medicine in Dayton, Ohio; associate professor of physiology, Howard University School of Medicine, Washington, D.C.; and clinical professor of medicine at Stanford University School of Medicine.

I have mentioned that Dr. Sandler is responsible for the overall management of NASA's Cardiovascular Research Office, which studies the hazards of aerospace flight with respect to the human body and its ability to adjust physiologically to the stresses of flight. The Office's work includes the study of not only phys-iological parameters, but also those of a biochemical nature including endo-crinology. In this area, Dr. Sandler has had the collaboration of Joan Vernikos, Ph.D. Her experience and expertise not only have been important in their own right, but also have added a new dimension to the Cardiovascular Research Office. The research areas that Dr. Vernikos has made her own include endo-crine pharmacology, mechanisms regulating the pituitary-adrenal system, stress and stress response, drug–stress interactions, effects of hormones on the central nervous system, space physiology and pharmacology, neuropeptides and pain, and circadian rhythms.

Dr. Vernikos received a bachelor's degree from the University of Alexandria in Egypt in 1955 and a Ph.D. from the Royal Free Hospital School of Medicine, University of London, in 1960. She was assistant professor of pharmacology at the University of Ohio in the 1960s, and she has been a research scientist at the NASA-Ames Research Center. During 1973–1978, Dr. Vernikos was chief, Human Studies Branch of the Biochemical Research Division at NASA-Ames, and for six months in 1976 she was also acting deputy director, Life Sciences. She is a lecturer in Aeronautics and Astronautics at Stanford University and is also a consultant in the Department of Psychiatry. She has been awarded the

NASA medal for exceptional scientific achievement. She is also a consultant for the European Space Agency, Paris, and the German Space Agency, Cologne, West Germany. Dr. Vernikos has had considerable editorial experience on the editorial boards of a number of scientific journals which, with her scientific expertise, makes her an ideal editor for the current volume.

The work of Dr. Vernikos has been an important complement to Dr. Sandler's studies, and this complementarity also makes them ideal coeditors for the current volume. Their intimate knowledge of their fields has enabled them to gather a distinguished group of contributors, and there is no doubt that this book will be among the most important and most comprehensive works on bed rest as well as a classic on this subject.

Geoffrey Bourne, D. Phil., D. SC., F.Inst. Biol., Fellow Roy. Soc. Med.
Professor of Nutrition and Vice Chancellor
St. George University School of Medicine
Grenada, West Indies

Preface

The flight of Sputnik across the sky on October 4, 1957, catapulted the world into the Space Age and resulted in a vast research program to determine the potential physiological effects that the removal of gravity (weightlessness) would have on humans. This research provided great masses of data not only on what could be expected in space, but also on the results of inactivity and immobilization on earth since these conditions afforded the best method we had to simulate—although not duplicate—the removal of gravity.

This book has been prepared to review the body of information from studies on healthy volunteers conducted in direct support of the space program and to discuss the present state of our knowledge of the physiological deconditioning inherent in inactivity and immobilization in general. It deals, in fact, with that end of the spectrum that is diametrically opposed to the growing literature on the benefits of exercise. It covers the changes that occur in the cardiovascular system, bone and muscle, metabolism and endocrine responses, psychosocial responses, and exercise tolerance. We have stressed clinical effects and clinical management of deterioration while indicating the changes that have been found in healthy, normal bed rested subjects. Where relevant, data on crews that have flown in space may also be included since their physiological responses are qualitatively similar to those observed in bed rested subjects or immobilized patients on earth.

Of primary importance is the development of means of counteracting the many physiological changes brought about by inactivity. As this book will show, few reliable countermeasures are currently available. This is one of the many areas where future research is greatly needed. Others are discussed under the individual subject areas. It is also important that we determine (a) whether the observed changes are due solely to physical inactivity or are compounded by the associated isolation, (b) whether they are reversible, and (c) whether they may result in serious health problems over the long term. Consequently, this book is directed primarily toward research investigators and students interested in this field in the hope of stimulating them to search for answers to the questions that still remain unanswered. The book should also provide useful information for clinicians and nursing staff in the management of their immobilized patients. Finally, it should offer a basic understanding of the physiology of inactivity for those in allied health disciplines who are concerned with the health problems and care of the sedentary and the aging.

No book like this could be written without help and assistance. We wish to thank the great number of co-investigators and collaborators both within the Space Agency and at various universities and research centers across the United States and abroad who contributed so unselfishly to this cause. We would like also to acknowledge the significant efforts of our Soviet colleagues who have shared invaluable information, and particularly those of A. I. Grigoriev and V. M. Mikhaylov of the Institute of Biomedical Problems, Moscow, who in 1979 worked so diligently to accomplish with us the first joint U.S.–U.S.S.R. bed rest study. Above all, we wish to express to all the healthy volunteers who participated in our studies our enormous appreciation for their cooperation and their willingness to adhere to schedules and strictly controlled experimental conditions. Without them, none of this information would have been available today. We would like to express our particular deep gratitude to two individuals without whose help we could not have prepared this material: first, to Ms. Mary Phares for her assistance in collecting background materials and in drafting and editing the various chapters, and then to Mrs. Doris M. Furman for the typing of the preliminary and final versions of each manuscript.

Harold Sandler
Joan Vernikos

Moffett Field, California

Inactivity: Physiological Effects

1

Introduction

**HAROLD SANDLER
AND JOAN VERNIKOS**
*Cardiovascular Research Office
Biomedical Research Division
National Aeronautics and Space Division
Ames Research Center
Moffett Field, California 94035*

Before the nineteenth century, sick people rarely took to their beds until they were too weak to stand or sit. This aversion to bed rest resulted primarily from the fear among working people of losing income critical to family survival and the general superstition that if you went to bed you would never get up again. All of this changed in 1863 when John Hilton concluded that if immobilization could heal a broken limb, it should be helpful in treating other health problems as well (Hilton, 1863). With lack of drugs and modern treatment modalities, the physicians of the time rapidly adopted this concept with so much zeal that they often put to bed many patients who would have been better off up and about, occasionally even forgetting some of them in the process. Starting 50 years ago, certain physicians began to question the prescribed practices then current of 4–6 weeks of bed rest following surgery or treatment of myocardial infarction and pointed out the clinical risks of prolonged bed rest (Asher, 1957; Browse, 1965; Thompson, 1934). Much of this was supported and further confirmed by experiences and findings from World War II, when men who were rapidly returned to ambulation had fewer problems than those bed rested for prolonged periods (Browse, 1965). Yet an understanding of these implications took time, a major reason being that all observations were made on sick people, whose basic illness was the cause of their going to bed. These illnesses could, and did, complicate observed findings.

The first studies to evaluate the effects of bed rest were conducted in the 1920s. The pioneers in the field were Campbell and Webster in 1921 and Cuthbertson in 1929. The first two investigators studied day and night cycles of nitrogen excretion in a 28-year-old male. Cuthbertson had a larger number of

1

subjects—five males (19–40 years) and two females (19–37 years). This investigator also studied the metabolic effects of immobilization. After these studies, little was done until the late 1940s when the question of whether or not immobilization would have deleterious physiological effects led to two studies (Deitrick et al., 1948; Taylor et al., 1949) that marked the beginning of the modern era of bed rest investigations, which continues to this day.

I. ENTRY INTO THE SPACE AGE

The advent of the space program provided yet another and important contribution. As humans moved to extend their presence in space from hours to days to months, methods for long-term simulation were sought in order to predict adverse reactions and to test and to evaluate various methods for their prevention. It was hypothesized, and later proven accurate, that physiological changes during spaceflight would be similar to those seen after bed rest. The space program provided the justification and resources to study the time course of these changes in large numbers of otherwise completely normal individuals confined to bed for prolonged periods. Rasults showed that all physiological systems of the body change with inactivity and immobilization. Some of these changes appear more rapidly or are longer lasting than others and will be described later in some detail. The severity of many of the changes depends on how long the individual is bed rested or inactive.

After bed rest, or spaceflight, individuals must again adapt to an erect body posture and to earth's gravity. This is often not so readily achieved as is adaptation to the state of inactivity or weightlessness. The return of affected physiological systems to normal can be highly variable and can require days, weeks, or even months. Moreover, certain individuals readapt more readily than others, and why this is so has not yet been determined. Changes in the musculoskeletal system continue to serve as an area for major concern to ground and aerospace medical investigations, because it is uncertain whether the observed changes are ever entirely reversed.

Finally, bed rest has proven to be an important physiological tool. Reactions have now been documented for almost 2000 normal adults and have included individuals up to age 65 years, athletes and nonathletes, and subjects from all walks of life (Greenleaf et al., 1982; Sandler, 1980; Nicogossian et al., 1979). These observations have proven invaluable in attempts to compare the system changes occurring in bed rested subjects with major clinical disease states (Browse, 1965; Steinberg, 1980). Similarly, the findings have also been important to the space program, since it has been difficult to obtain accurate and reproducible data in space. This has usually occurred for three reasons: first, mission operational requirements and crew activities have taken precedence over medical evaluations; second, since the health and safety of crew members are of primary concern, countermeasures are often employed, some of which alter physiological responses; finally, the relatively small number of astronauts and cosmonauts who have flown in space makes it difficult to extrapolate the data to reach general conclusions about the health and safety of humans during very long exposures to weightlessness. To date, a little more than 300 individuals (9 of whom were women) have flown in space; however, 210 of these persons have been in space for 14 days or less, and only eight, all Soviets, have ever flown 6 months or longer. To overcome these problems, both U.S. and U.S.S.R. investigators continue to rely on ground-based methods, such as bed rest, to obtain statistically significant information on effects and to test the broadest possible segment of the population for change and reaction to proposed countermeasures.

II. GROUND-BASED SIMULATIONS OF WEIGHTLESSNESS

The methods used to simulate null gravity on earth include immersion, bed rest, chair rest, isolation, hyperbaric environments, and immobilization of animals (Sandler, 1980). None of these techniques precisely duplicate weightlessness, because gravity cannot be entirely eliminated on earth. Immersion and bed rest studies, however, have provided possibilities for long-term exposures, with findings closest to those seen with a weightless state. Head-out immersion has the longest history of use (Epstein, 1978; Sandler, 1980). This technique has provided simulations lasting for several hours to as long as 56 days (Shulzhenko *et al.*, 1977). It has been used most extensively to document physiological adaptation to rapid shifts in body fluid distribution. Water immersion longer than 12 hr is difficult, because of problems with temperature control of immersion fluids, personal hygiene, negative-pressure breathing, skin maceration, and associated psychological problems from sensory deprivation. Consequently, bed rest, which does not have these limitations, has been used much more frequently.

More than 160 bed rest studies using healthy individuals have been conducted by U.S., Soviet, and other European investigators to evaluate physiological changes occurring in weightlessness (Greenleaf *et al.*, 1982; Nicogossian *et al.*, 1979; Sandler, 1980). The greatest percentage of studies have investigated cardiovascular changes; those studying exercise tolerance, possible countermeasures, hormonal and metabolic changes, bone mineral losses, and muscle atrophy and neurophysiological changes are the next most common. Subjects have been stress-tested following bed rest with 70° head-up tilt, lower-body negative pressure, centrifugation, and exercise to determine decrements in work performance.

Toward the end of the 1960s, Soviet investigators evolved a new method of bed rest in which subjects were positioned with the head lower than the feet, rather than horizontal, after subjective comments of cosmonauts after flight suggested that the head-down position more closely reproduced the feelings of head fullness and awareness experienced during flight. The first study (Genin *et al.*, 1969) compared responses from horizontal bed rest and −4° head-down tilt. Since then additional studies have been conducted with head-down positions ranging from −2° to −15° and lasting from 24 hr to 182 days (Kakurin, 1981; Kakurin *et al.*, 1976; Katkov *et al.*, 1982; Sandler, 1980). In general, head-down bed rest induces findings more rapidly and profoundly than its horizontal counterpart.

Both inactivity and removal of postural stimuli, associated with a change in the direction of gravity pull on the body, are suspected to cause the physiological changes seen in spaceflight and immobilization. In both spaceflight and immobilization, these factors are closely interrelated. In space, the body becomes weightless, with loss of load bearing on bone and muscle, particularly the legs, and there is no longer a gravity gradient from the upper to lower body segments on becoming erect. Furthermore, the body cannot move about as readily as it did on earth and essentially becomes inactive. With immobilization, gravity is still present, but head-to-foot loading of the body along its long axis is minimized. Shifting of body fluids footward cannot take place. Loss of these postural cues in bed rest is therefore one of its most important conditioning factors towards subsequent altered physiological reactions. Under such hypodynamic conditions metabolic demands are severely reduced, since the body has had both its type and range of motion restricted, as compared to when it is upright and mobile. Normal basal endocrine and neurophysiological functions require periodic stimulation by such alternating and recurring cues as meals, light and dark cycles, noise, and changes in position. These functions must receive particular and careful attention during either spaceflight or bed rest; otherwise, they can also become factors causing physiological changes.

III. EFFECTS OF INACTIVITY

The human body is constructed for movement, as evidenced by the fact that skeletal muscle constitutes 40% of body mass. During strenuous exercise, blood flow in muscle can be 15–20 times greater than during rest (Steinberg, 1980). The metabolic rate of muscle at peak exercise can be 50–100 times greater than during supine rest (Steinberg, 1980).

The circulation and respiration provide the working muscle with nutrients and oxygen while removing metabolic waste and carbon dioxide. These two systems respond efficiently to any increase in muscular activity. Cardiac output, heart rate, and left ventricular function all increase. Even with moderate exercise, cardiac output may triple, heart rate may double, and left ventricular effort may more than triple. Oxygen uptake with heavy exercise rises 6 times higher than seen at rest, and minute volume of respiration may increase 10 times over resting values. With maximal exercise, tidal volumes may reach as high as 50% of vital capacity. These changes are greatest in physically fit individuals and are severely reduced in inactive or immobilized persons. With inactivity, physical fitness decreases rapidly, and maximal cardiorespiratory response, a measure of reserve capacity, is severely reduced. Finally, orthostatic intolerance occurs in almost all individuals.

Body metabolic function level is significantly affected by inactivity, although decreases in myocardial and postural muscle work are related to gravity (Browse, 1965; Vernikos-Danellis *et al.*, 1974). General metabolism declines, including caloric and dietary requirements, as part of an adaptive process responding to a new physiologic state.

The efficiency of many other systems also declines with immobilization. Human performance shows decrements over the long term. The intellect is dulled and a state of apathy and depression ensues. Changes are complicated by the associated isolation and confinement, with attendant disassociation from family and friends. Prolonged inactivity also impairs motor ability and task precision. This change occurs because of altered sensory perception and functional mainte-nance of some muscles and atrophy of others, leading to changes in abilities requiring precise coordination. The immobilized individual loses the ability to coordinate his or her movements rapidly and efficiently.

During prolonged immobilization, the gastrointestional tract is also affected. Immobilized individuals lose their appetite, often for protein-rich nutrients that would contribute to muscle building. Renal problems also occur, because the body is not in a satisfactory position to totally expel urine, and calculi may form in the stagnant pools (bladder or kidney hila) from low-level, long-term increases in urinary calcium content. Prolonged inactivity, in fact, affects every organ in the body, disturbs hormonal and metabolic functions, and contributes to bone mineral loss and osteoporosis. Finally, immobilized individuals cannot respond effectively to stress, and psychosocial interactions may suffer severely.

IV. THE EFFECTS OF GRAVITY

Humans are affected by gravity more than any other animal because of their normal erect posture. Over time the ability to bear weight on the spine and walk upright has required significant adaptive changes. The use of the legs to walk and to return the shifted and reasonably large volumes of blood from the lower extremities has required unique physiological modifications. All such changes are no longer needed or useful in a weightless environment. Let us consider in some detail how gravity affects various physiological systems.

A. Bone and Muscle

Ever since humans attempted to stand upright, gravity has tried to pull them down. The compressive force of gravity on the spinal column and leg joints is such that each astronaut or cosmonaut has reported an increase in body length of at least 2–5 cm during flight. Standing is an active process and results in a small, but significant, expenditure of energy (at least 5.71 kcal/min, 2/hr), which is a 16–19% increase compared to lying down (Browse, 1965). Antigravity muscles come into play and primarily consist of the quadriceps femoris, gluteii, and spinal erectors; the actions of these muscles are balanced by the hamstrings and anterior abdominal muscles. Response and interactions are under central nervous system control, particularly guided by proprioceptive inputs from fibers and receptors in joints and muscles and spinal–cerebellar–vestibular influences. Absence of weight bearing not only removes the direct compressive force on the long bones and spine but also removes the indirect loading on these bones from the pull of these muscles on the various bony structures to which they are attached. Unloading of the skeleton invariably leads to osteoporosis, a weakening of bone strength, and delayed ability to heal fractures, should they occur. Skeletal muscle, as indicated, also makes up 40% of body mass. Disuse and inactivity lead to adaptive changes resulting in change in both size and fiber type for individual groups of muscles, particularly those involved in weight bearing. The underlying reasons for changes in bone and muscle are not known, but the debilitation continues unabated throughout the period of altered gravity if attempts at prevention are not employed. It remains unclear at present whether the changes are irreversible. To date, osteoporotic changes in postmenopausal women and the elderly have generally been refractory to treatment (Steinberg, 1980).

B. The Circulation

Gravity produces significant effects on the circulation and the need for adaptive change. In the upright position, with gravity in full effect, at least 10–15% of the circulating blood volume is shifted footward. Yet such a shift is accomplished with little change in blood pressure or oxygen delivery to the brain. This is the result of the close interaction between the central nervous system (baroreceptors) and the heart (rate and stroke volume). Reactions also include control of venous storage capacity of the lower extremities and supplemental information released by the peripheral tissues as hormones (adrenal and renal steroids) and/or metabolic end products (prostaglandins, adenosine 5'-monophosphate, lactate) that indicate the adequacy of oxygenation. Since we usually spend at least two-thirds of our time erect or sitting, these processes of adaptation and regulation for the headward return of gravity-displaced body fluid have become important steps in our evolutionary process. During weightlessness, however, such mechanisms are not needed, and in-flight evidence shows that the body once again readily adapts to this new physiological state and does so successfully. Changes during flight and inactivity occur rapidly, and the ability to adjust to an upright body position (lower-body negative pressure [LBNP], stand test, or 70° head-up tilt) are reduced. Using careful hemodynamic and neurohumoral measurements, significant changes can be shown after only 6–8 hr of bed rest, 1–2 hr of water immersion, or 1–2 days of weightlessness. Continued exposure increases both the severity of response and the time needed for recovery. Loss of orthostatic response when used to measure physiological change continues to decrease during the first 3–4 weeks of bed rest or spaceflight and shows much slower, or little, change after the fifth to sixth week, as a state of full physiological adaptation takes place to the new condition of inactivity.

C. Renal and Fluid/Electrolyte Function

Withdrawal of gravity, such as occurs with a supine body position or weightless state, is associated with a significant headward shift of body fluid and sets in motion neurohumoral and central nervous system changes that attempt to handle the apparent excess. Acute hemodynamic changes involve increases in renal pressure and flow. These are followed quickly by neural and hormonal responses designed to increase renal free water clearance (through a decrease in antidiuretic hormone release) and eventually to increase salt and potassium excretion. Renin-angiotensin excretion is also suppressed. Long-term exposure to bed rest of at least several months duration is associated with suppressed adrenal responsiveness and suppressed ability to cope with all forms of physiological stress. Urinary excretion of calcium rises slowly and remains persistently elevated at a low level after the first few weeks of exposure. Such elevated urinary calcium levels have proven troublesome, leading to stone formation in cases with urinary tract abnormalities, particularly in the presence of infection.

D. Sleep and Immobilization

It has been suggested that changes during prolonged bed rest are no more serious than those seen during sleep. However, that is not so. During night sleep in a normal, healthy human, heart rate decreases, blood pressure falls, and peripheral dilation occurs in skin blood vessels, resulting in a secondary loss of body heat. Plasma protein concentration and hematocrit fall, and fibrinolytic activity is decreased so that blood clots form more readily during sleep. Respiration becomes slower and shallower as sleep deepens. In deep sleep, alveolar and total ventilation decreases, and there may be periods of apnea and Cheyne-Stokes respiration. Although the alveolar oxygen concentration remains unchanged, alveolar carbon dioxide concentration increases. The latter change, however, affects the respiratory system very little. Unlike bed rest, sleep decreases urine production, as well as the excretion of electrolytes, corticosteroids, and hormone breakdown products. The central nervous system shows a decrease in the amplitude of reflex responses resulting from decreased cerebral activity. Although cerebral blood flow may increase to compensate for any fall in blood pressure, cerebral oxygen consumption remains unchanged.

V. SUMMARY

The findings of research personnel and clinicians have demonstrated that no single human physiological system remains unchanged following prolonged periods of inactivity or immobilization. Under these conditions, the various systems deteriorate to a lesser or greater degree and according to different time sequences. Although we have accomplished a vast amount of information on the physiological responses of individuals to these conditions, numerous questions remain to be answered as to why these changes take place, whether they are reversible, and, most importantly, whether they will result in health problems many years later as the individual ages. Research is being conducted at present to find some of the answers. But many comprehensive studies will be needed in the future if we are (a) to understand fully the implications of inactivity and immobilization for research subjects, clinically bed rested or paralyzed patients, and the elderly on earth, as well as for long-term space station dwellers, and (b) to develop countermeasures to offset the observed physiological responses.

REFERENCES

Asher, R.A. (1957). *Br. Med. J.* **2**, 967.
Browse, N.L. (1965). "The Physiology and Pathology of Bed Rest." Charles C. Thomas, Springfield, Illinois.
Campbell, J.A. and Webster, P.A. (1921). *Biochem. J.* **15**, 660-664.
Cuthbertson, D.P. (1929). *Biochem. J.* **23**, 1328-1345.
Deitrick, J.E., Whedon, G.D., and Shorr, E. (1948). *Am. J. Med.* **4**, 3-36.
Epstein, M. (1978). *Physiol. Rev.* **58**, 529-581.
Genin, A.M., Sorokin, P.A., Gurvich, G.I., Dzhamgarov, T.T., Panov, A.G., Ivanov, I.I., and Pestov, I.D. (1969). *In* "Problems of Space Biology" (A.M. Genin and P.A. Sorodin, eds.), Vol. 13, pp. 256-262. Nauka Press, Moscow.
Greenleaf, J.E., Silverstein, L., Bliss, J., Langenheim, V., Rossow, H., and Chao, C. (1982). *In* "Physiological Responses to Prolonged Bedrest in Man: A Compendium of Research", pp. 1-115. NASA TM-81324, National Aeronautics and Space Administration, Washington, DC.
Hilton, J. (1863). "Rest and Pain." (Edition by E.W. Walls and E.E. Phillips, 1953). G. Bell and Sons, London.
Kakurin, L.I. (1981). *In* "12th US/USSR Joint Working Group Meeting on Space Biology and Medicine," pp. 1-36. Washington, DC.

Kakurin, L.I., Lobachik, V.I., Mikhaylov, M., and Senkevich, Y.A. (1976). *Aviat. Space Environ. Med.* **47**, 1084-1086.

Katkov, V.Ye., Chestukhin, V.V., Nikolayenko, E.M., Grozdev, S.V., Rumyantsev, V.V., Guseynova, T.M., and Yegorova, I.A. (1982). *Space Biol. and Aerosp. Med.* **16(5)**, 64-72.

Nicogossian, A.E. and Parker, J.F., Jr. (1982). "Space Physiology and Medicine," pp. 1-324. NASA SP-447, National Aeronautics and Space Administration, Washington, DC.

Nicogossian, A.E., Sandler, H., Whyte, A.A., Leach, C.A., and Rambaut, P.C. (1979). "Chronological Summaries of United States, European, and Soviet Bedrest Studies," pp. 1-370. Biotechnology Inc. and Office of Life Sciences, National Aeronautics and Space Administration, Washington, DC.

Sandler, H. (1980). *In* "Hearts and Heart-Like Organs", (G.H. Bourne, ed.), Vol. 2, pp. 435-524. Academic Press, New York.

Shulzhenko, E.B., Vil'vilyams, I.F., Grigoriev, A.I., Goglev, K.I., and Khudyakova, M.A. (1977). *In* "Life Sciences and Space Research," (P.H.A. Sneath, ed.), Vol. XV, pp. 219-224, Pergamon Press, New York.

Steinberg, F.U. (1980). "The Immobilized Patient: Functional Pathology and Management," pp. 1-156. Plenum Press, New York and London.

Taylor, H.L. Henschel, J., Brozek, J., and Keys, A. (1949). *J. Appl. Physiol.* **2**, 223.

Thompson, T.S. (1934). *J. Bone Joint Surg.* **16**, 564.

Vernikos-Danellis, J., Winget, C.M., Leach, C.S., and Rambaut, P.C. (1974). 'Circadian, Endocrine, and Metabolic Effects of Prolonged Bedrest: Two 56-Day Bedrest Studies," NASA TMX-3051, pp. 1-45. National Aeronautics and Space Administration, Washington, DC.

2

Cardiovascular Effects
of Inactivity

HAROLD SANDLER

Cardiovascular Research Office
Biomedical Research Division
National Aeronautics and Space Administration
Ames Research Center
Moffett Field, California 94035

I. THE IMPACT OF GRAVITY

Gravity is a consistent factor governing life on earth. As humans, we normally spend two-thirds of our day standing or seated. While erect, a significant amount of intravascular volume and tissue fluids are shifted to the lower body from the pull of gravity, and the body must compensate to maintain blood flow to the head and to distribute blood volume adequately throughout the body. When the compensatory mechanisms are inadequate or retarded, orthostatic intolerance or hypotension occurs, with eventual fainting.

Over time, the human body has evolved a gravity receptor system that uses information gained from muscle proprioceptors, semicircular canals, otoliths, and mechanoreceptors (baroreceptors). These systems, which sense body position and initiate the necessary change, operate continually as we interact with our environment and are most evident when we change from the supine to erect body position. Normally, 70% of the body's blood volume resides in systemic veins, 15% in the heart and lungs, 10% in systemic arteries, and 5% in capillaries. The upright position shifts 700 ml of venous blood from the upper body to the legs, with 400 ml coming from the central circulation (heart and lungs). As shown in Fig. 1, the loss in central blood volume immediately causes a 25% decrease in cardiac output, a 25% increase in heart rate, and a 40% decrease in stroke volume, with little change, or even a slight increase, in blood pressure. Blood pressure is maintained by an increase in flow resistance through arterial and arteriolar constriction, resulting from increased sympathetic nervous system out-

INACTIVITY: PHYSIOLOGICAL EFFECTS

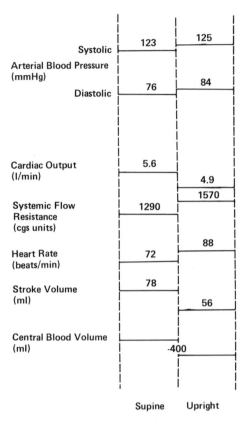

Fig. 1. First level of defense against gravity: hemodynamic changes from supine to upright posture.

put, due to triggering of aortic and carotid sinus baroreceptors. The process is reversed when the body changes from erect to supine.

II. LEVELS OF DEFENSE IN BODY POSITION CHANGES

There are at least three levels of defense to offset cardiovascular changes that occur with changes in body position.

The first is an adjustment in venous capacity and pressure by redistributing contained volume. The magnitude of the venous shift is shown schematically in Fig. 1. When erect, shifted volume is transferred primarily to deep intra- and intermuscular leg veins, with about 200 ml going to the pelvis and gluteus maximus areas. Adjustments for the displaced venous volume occur and rely on the rapid contraction and relaxation of smooth muscle in the venous wall, respiration (movement of the diaphragm), and particularly, contraction of the lower limb muscles, which by their action squeeze blood back toward the heart.

The second line of defense occurs when the pre- and postcapillary sphincters act to increase or to decrease fluid in the tissues. As shown in Fig. 2, after 10 min of standing this mechanism shifts about 10% the of plasma volume to dependent tissue spaces; such loss stabilizes at about 15% after 20 min. This second line of defense is probably more important in the long run than are fluid shifts within venous vessels, because it increases or decreases the absolute volume of the cardiovascular system, as opposed to redistributing its contents. Factors that can affect this ability for volume change occur at the local tissue level and include nervous, metabolic, or biochemical events.

Finally, there is a third line of defense, which is under neurohumoral control and is used primarily for long-term adjustments. The events associated with these changes are detailed in Fig. 3. Activation of neurohumoral control depends on stimulation of proprioceptors (baroreceptors) which, depending on their design and whether they are located inside or outside the chest, are able to sense (a) the filling of the system, (b) the wall tension in atria or ventricles, or both, and (c) the pulse and mean pressures in the pulmonary and systemic arteries. When activated, the system stimulates sympathetic nerve activity or inhibits the release

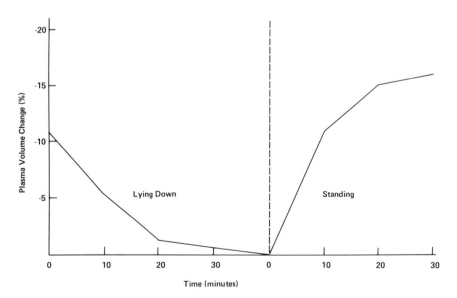

Fig. 2.   Second level of defense against gravity: plasma volume changes during lying down and standing. [From Hagan *et al.*, 1978.]

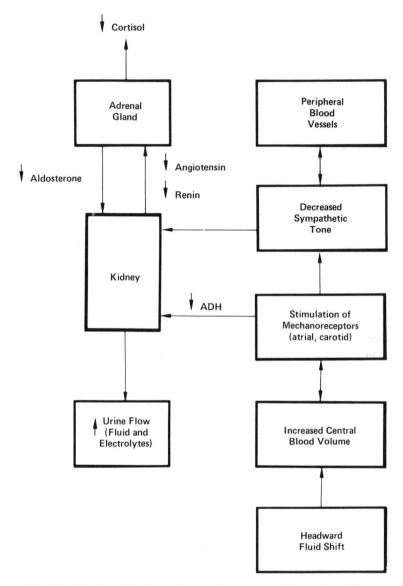

Fig. 3. Third level of defense against gravity: responses in the supine position.

of potent hormone regulators, such as antidiuretic hormone (ADH)(Gauer and Henry, 1976). Long-term excitation of baroreceptors also changes the secretion of adrenal hormones (e.g., aldosterone) and the concentration of pressor substances (e.g., renin-angiotensin) from the kidneys. Responses can be measured during changes in environmental temperature or altitude; during exercise; under certain disease conditions, such as diabetes insipidus; or with inappropriate secretion of ADH. Studies to date suggest that this last means of controlling blood volume is the one used by the body for long-term adaptation of the circulation to gravitational changes. Recently, it has been suggested that ADH may play a role in maintaining proper blood pressure through regulating total circulating plasma volume (Padfield and Morton, 1983). It has also been clearly shown that atrial distention may release a natriuretic factor that controls loss of sodium and other electrolytes by the kidneys (deBold, 1982; Dewardener and Clarkson, 1982; Thibault et al., 1983).

The incidence of fainting and its rate of onset in humans can be elicited during standing or tilting by blood withdrawal, the use of vasodilator drugs, immobilization, lower-body negative pressure (LBNP), or the use of venous-occlusive tourniquets or cuffs. The cause appears to be an increased vagal discharge superimposed on the already-basic sympathetic response to tilt or change in body position and, therefore, has been termed "vasodepressor or vasovagal syncope" (Sandler, 1977). The exact trigger mechanism for the vagotonic discharge remains unknown, but the work of Epstein and co-workers (1968a,b) strongly suggests a primary role for the cardiac ventricles, which, upon a shift of intravascular volume to the feet, reach a critical minimal cardiac size. At this point sufficient nerve fibers in the wall become stimulated by the ventricular contraction to initiate the nerve discharge required to start the fainting process. Such a possibility in bed rested subjects or athletes is supported by the results of Sandler and co-workers (Sandler et al., 1977; Sandler and Winter, 1978).

Vasovagal syncope during tilt is characterized by the sudden onset of bradycardia, nausea, pallor, sweating, dizziness, and an abrupt drop in blood pressure. In general, such syncopal reactions occur in any clinical state that leads to a sudden reduction of heart rate, stroke volume, or peripheral resistance (Sandler, 1977). Severe postural hypotension regularly occurs in a variety of clinical conditions, such as diabetes and Addison's disease, or with other more uncommon states that disturb peripheral sympathetic innervation—for example, idiopathic chronic orthostatic hypotension (Ibrahim et al., 1975). Investigators of this last condition (Ibrahim et al., 1975; Ziegler et al., 1977) have shown that such subjects do not pool more blood in dependent regions than do normal individuals, but the resultant pooling nevertheless causes a marked fall in blood pressure. Changes are due to the absence or deficiency of the first two defense mechanisms previously described, since they are unable to respond with reflex arteriolar and venous constriction to the imposed volume load. These findings are the opposite of changes occurring during a vasovagal faint. The responses after prolonged bed rest and water immersion represent a mixture of losses of all three defenses, since they are associated with distinct changes in body fluid volumes (intra- and extravascular) and volume distribution, as well as deterioration of the normal reflex mechanisms that are responsible for peripheral vascular tone. The conditions appear normal, or compensated for, as long as the body remains horizontal. The changes are clinically manifested primarily when individuals stand up suddenly or are stressed by upright tilting or LBNP. The resulting postural hypotension is regularly accompanied by decreased pulse pressure and tachycardia. In more susceptible individuals, when orthostatic stress is continued, vasodepressor syncope occurs.

III. THE GAUER-HENRY REFLEX

The Gauer-Henry reflex hypothesis, which has long been accepted, has been claimed as the main explanation for the manner of initiation of events leading to long-term fluid regulation, as shown in Fig. 3. In this way, distention of atrial receptors (mechanoreceptors), resulting from increased central blood volume, as in going from erect to supine, signal to the central nervous system, through the vagus nerve, that there is excessive fluid present; this leads to ADH suppression (Gauer et al., 1970; Goetz et al., 1975). The result is a subsequent diuresis, or loss of fluid by the kidney. This hypothesis was based on data from dogs, in which atrial distention has resulted in diuresis in the majority of animals but is prevented by vagus nerve section. Recent work with nonhuman primates, however, casts doubt on this explanation, since vagotomy in this animal model has had no effect in preventing the diuresis caused by head-out water immersion (Gilmore, 1983). Such findings suggest that atrial stretch receptors are far less sensitive in primates, and possibly humans, and lead to the conclusion that high-pressure baroreceptors, rather than low-pressure atrial receptors, are responsible for the control of urinary responses in these cases. The possibility that this may occur in humans has been demonstrated recently during tilt and water immersion studies of heart and heart–lung transplant recipients (Convertino et al., 1983). These subjects, who have limited nerve (vagal) connections to the heart, should not have shown plasma volume, plasma renin activity (PRA), and ADH responses of normally innervated individuals with an intact vagus nerve; yet they did. This observation is highly significant, since the regulation of body fluids and electrolytes is one of the chief physiological systems responding to change in the state of activity on earth or in spaceflight.

IV. CARDIOVASCULAR RESPONSES TO INACTIVITY AND IMMOBILIZATION

Bed rest has been the method used most frequently for simulating weightlessness in humans. A great number of healthy individuals have therefore been studied, as shown in Table I, with durations ranging from several hours to over 7 months. The majority have been younger males, 19–35 years, but more recently, normal older individuals, up to 65 years, have also been tested and have included both females and males (Goldwater *et al.*, 1981; Sandler *et al.*, 1979).

In contrast to normal activity, body position during bed rest is maintained horizontal (or head-down) without a break for days, weeks, or months. Furthermore, all emunctory functions are carried out while the subject is horizontal or head-down. The subjects are allowed to rise up only on one elbow to eat for a period of 20–30 min three times a day. In some studies a pillow is not allowed, nor is lifting of the head for any reason. Reading is accomplished using prism glasses or by appropriate elevation of the reading material. Studies have usually been carried out in two- to eight-bed wards, with the subjects entertained by television, radio, and visitation by the facility staff; but, in general, the subjects are isolated from home, friends, and the outside world.

The cardiovascular changes occurring with bed rest are summarized in Table II. A diuresis over the first few days of exposure is a universal finding and is associated with a rapid, 8–10% loss of plasma volume, which tends to stabilize at a 15–20% decrease after 2–4 weeks (Greenleaf, 1983). The mechanisms for this loss are shown in Fig. 3 and are primarily associated with a suppression of ADH secretion. These fluid volume changes during bed rest parallel findings with head-out water immersion but are of lesser magnitude (Epstein, 1978). If bed rest is continued, there is a continued slow loss of plasma volume; bed rest studies lasting 100–200 days have registered 30% losses (Greenleaf, 1983). Concurrent with this loss, the hematocrit increases, and when bed rest continues for longer than 2 weeks, it leads to a decrease in red cell mass and a slight reticulocytosis (Burkovskaya *et al.*, 1980; Morse, 1967). These latter changes level off after 60 days of bed rest, remaining stable but depressed for the rest of the bed rest period (Burkovskaya *et al.*, 1980). The stabilizing influence is probably due to an increase in bone marrow activity, but the exact causes are still unknown (P. C. Johnson *et al.*, 1977). Similar hematologic findings have occurred with spaceflight (Kimsey, 1977). Echocardiographic measurements of heart size parallel changes in plasma volume and are influenced by the state of athletic conditioning of the subject, with athletes showing expected larger resting values and greater bed rest losses. End-diastolic volume after 2 weeks of horizontal bed rest in seven athletically trained male subjects, 19–25 years old, fell 11% ($p < .01$) from $70 \pm 2 \text{ ml/m}^2$ to $62 \pm 3 \text{ ml/m}^2$. In seven similarly aged nonathletic males,

TABLE I

Representative Horizontal Bed Rest Studies[a]

Duration	No. of studies	No. of subjects[b]	Representative study areas
30 min to 48 hr	15	212 (16 F)	Fluid volume redistribution; effect of LBNP on orthostatic tolerance; comparison of responses to chair rest, bed rest, and immersion; exercise and occlusive cuffs as countermeasures
3–7 days	31	305	Reactive hyperemia in forearm and calf muscles; exercise, LBNP, and fluid injestion as countermeasures
8–14 days	44	448 (32 F)	Central and peripheral circulatory function; O_2 alveolar tension changes; leg volume changes; plasma responses in females during menstrual cycle; drugs, fluid injection, and exercise as countermeasures
15–21 days	13	261 (8 F)	Cardiac size and function; heart rate and cardiac output with exercise stress; treadmill and trampoline exercise and restrictive clothing as countermeasures
24–28 days	13	187 (1 F)	Metabolic effects of bed rest; changes in blood volume; fluid and electrolyte responses; cardiac catheterization testing; antigravity suits and exercise as countermeasures
30–35 days	10	75	Plasma volume responses with tile; red blood cell kinetics; physical training and exercise as countermeasures
40–49 days	10	151	Adjustment of cardiovascular system to physical work; physiologic and metabolic functions; exercise, electrostimulation of muscles, and occlusive cuffs as countermeasures
56–75 days	6	75	Cardiac, central nervous system, and hormonal activity; exercise and drugs as countermeasures
84–560 days	9	88	Cardiac output measured by acetylene method; physiological responses; fluid and electrolyte shifts; exercise, drugs, and potassium phosphate as countermeasures

[a]Additional information on horizontal bed rest studies may be found in Greenleaf et al. (1982), Nicogossian et al. (1979), and Sandler (1980).
[b]F, females; otherwise, males.

TABLE II

Cardiovascular Changes with Bed Rest

Plasma volume loss of 15–20%
Total blood volume loss of 5–10%
Decrease in heart volume of 11% after 20 days of bed rest
Decrease in left ventricular end diastolic volume of 6–11%
No change or an increase in basal heart rate
No change or a decrease in supine cardiac output and stroke volume
Reduced orthostatic tolerance during standing, tilt, or LBNP
Reduced exercise tolerance; decreased $\dot{V}_{O_2max}$

---

end-diastolic volume changed from 62 ± 3 ml/m^2 to 58 ± 3 ml/m^2 (6%, $p < .05$) (Sandler et al., 1985b). A 12% decrease has been observed in nonathletic females (Sandler, 1980). Resting stroke volume and cardiac output also fell during bed rest, as would be predicted from the decrease in heart size and lowered metabolic demand associated with inactivity and loss of muscle mass. In the males, resting stroke volume fell 3–9% and cardiac output 6–13%, females had respective 25% and 21% decreases (Sandler, 1980). The decreases in end-diastolic volume observed in all cases parallel previously reported decreases in heart volume (both atrial and ventricular) seen with X-rays (Krasnykh, 1973; Saltin et al., 1968).

Bed-rest-induced changes in stroke volume, and consequently cardiac output, occur in several stages. The first stage takes place immediately after the body changes from upright to supine, as body fluids shift toward the head (see Fig. 1). Central circulating volume is expanded, leading to increased ventricular filling pressure and heart volume. These changes lead to an increase in stroke volume and cardiac output. Over the next 24–48 hr, as bed rest continues, the low-pressure baroreceptors and volume receptors react to restore the central volume overload to normal. Then, stroke volume and cardiac output decrease toward normal levels—changes that are consistent with the observed diuresis and accompanying decrease in heart size and end-diastolic volume. Inactivity or bed rest of very long duration (months or even years) results in a further adaptation in which stroke volume and cardiac output continue to decrease and eventually stabilize at significantly lower levels than normal (Kovalenko and Gurovskiy, 1980).

Resting systolic and diastolic blood pressures during bed rest generally remain within normal limits, although there is a tendency for diastolic values to rise (Polese et al., 1980). Heart rate, on the other hand, usually shows small increases with time. An early study by Taylor et al. (1949) found that heart rate increased by 0.5 beats per minute (bpm) each day of bed rest. Bed rest studies lasting up to 10 days have shown heart rate increases of 12–32 bpm over pre-bed-rest values. The increase appears to level off after 60 days of bed rest and to increase thereafter by 1–5 bpm per week. Although the majority of investigators have shown slight increases, others have reported little or no change in heart rate with bed rest (Kakurin *et al.*, 1980) or even a decrease (Kovalenko and Gurovskiy, 1980).

Electrocardiographic results have varied. In subjects bed rested for 10 weeks and monitored weekly by standard limb leads or vectorcardiography, there have been slight, but significant, changes in T-waves (Korolev, 1969). As the duration of bed rest lengthened, T-wave amplitudes measured by standard limb leads increased. Increases in U-wave amplitudes have been observed in precordial leads and have become significant by the second or third week of bed rest. Six months of bed rest produced increases in the QRS complex amplitude and decreases in the T-wave amplitude (Turbasov, 1980). Bed rest subjects have also exhibited significant increases in heart rate and decreases in T-wave amplitude when anticipating, or undergoing, venipuncture (Tkachev and Kul'kov, 1975). Anxiety, however, may not be the entire answer to these changes in wave form, since it is known that electrocardiographic (ECG) changes can be associated with increases and decreases in heart size, or shift of heart location in the chest, as well as altered hematocrit, all of which occur with bed rest (Nelson *et al.*, 1972).

The return to normal following a period of prolonged bed rest can be variable and can take a long time. Control of blood pressure during orthostatic provocation (5 min stand test, 15-min 70° head-up tilt or lower-body suction) usually returns to pre-bed-rest levels and stabilizes within 24–72 hr of ambulation, whereas the heart rate response may be exaggerated until the end of the first to third week of rehabilitation (Vetter *et al.*, 1971). X-ray measurements of heart size and stroke volume following 30–100 days of bed rest showed that neither had returned to normal, even after 60 days of recovery—and this despite a regular exercise program (Krasnykh, 1974, 1979). Systolic time intervals did not return to control levels for 4 weeks following 28 days of bed rest (Hyatt *et al.*, 1973).

Athletic state prior to the bed rest exposure also may serve as a complicating factor, since athletic subjects require a considerably longer period (3–4 weeks) than do sedentary subjects for such variables as heart size and maximum uptake of oxygen ($\dot{V}_{O_2\,max}$) to return to pre-bed-rest levels (Saltin *et al.*, 1968; Sandler, 1980). Athletic subjects exposed to three consecutive 2-week periods of horizontal bed rest showed incomplete recovery of heart rate and end-diastolic volume despite 3 weeks of ambulation between exposures and a program use of isotonic or isometric exercise during the course of the bed rest (Sandler *et al.*, 1985b). Figure 4 shows the heart rate and end- diastolic volume (measured by echocardiography) responses in these subjects at rest (control) and during LBNP. Resting (pre-LBNP) values of end-diastolic volume prior to the second and third bed rest periods showed the continued presence of deconditioning (lower levels), and the response to LBNP was more severe with each bed rest exposure. Heart rate

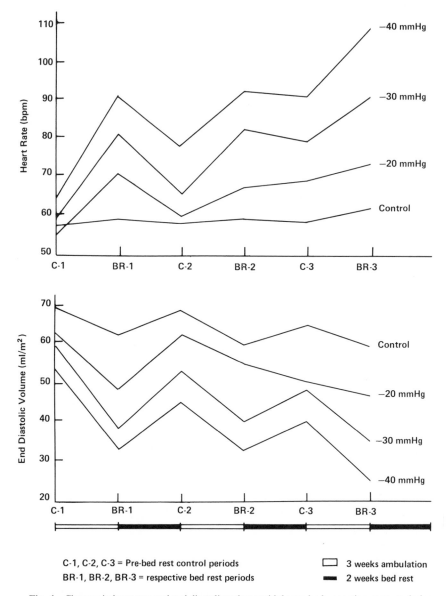

C-1, C-2, C-3 = Pre-bed rest control periods   ☐ 3 weeks ambulation
BR-1, BR-2, BR-3 = respective bed rest periods   ▬ 2 weeks bed rest

Fig. 4 Changes in heart rate and end diastolic volume with lower-body negative pressure during repeated 2-week bed rest exposures.

response to each LBNP test demonstrated progressively higher levels with each subsequent bed rest period, despite 3 weeks of full ambulation between bed rest intervals.

Recently, head-down bed rest has also been used to provide information on the cardiovascular changes that occur with inactivity and weightlessness. This technique was developed by Soviet investigators (Genin *et al.*, 1969; Genin and Kakurin, 1972) with body positions that have now ranged from $-2°$ to $-15°$ and for durations lasting from 24 hr to 182 days. Table III summarizes the fairly large number of studies conducted to date. Most head-down bed rest studies have used angles of $-4°$ and $-6°$, since these positions seem to best imitate responses to weightlessness (Kakurin *et al.*, 1976a,b). Subjective responses have included complaints of blood rushing to the head, heaviness of the head, and vessel pulsations in the temples (Blomqvist *et al.*, 1980). Objective findings have been neck vein engorgement, increased venous distention of the retinal veins, and increased central and jugular venous pressures. Most physiological responses to head-down bed rest reach maximum intensity within 3 hr after assuming the head-down position and mimic those experienced during the first few hours of spaceflight. As the head-down bed rest angle is increased, subjective and objective changes become increasingly severe.

Central venous pressure (right atrial) obtained during head-down bed rest is shown in Fig. 5 and is compared with findings during head-out immersion. Nixon and associates (1979) saw a sharp increase in central venous pressure during the first 40 min of $-5°$ head-down bed rest and a return to baseline levels by 90 min. After that, there was a continual gradual decline to a level significantly below baseline values by 12–16 hr of the 24-hr study. Similar changes were seen in a 7-day, $-15°$ head-down bed rest study (Katkov *et al.*, 1982). Mean right atrial pressure during this latter study did not change significantly over the first 7 hr but decreased to a little more than half-baseline values by the second day of bed rest and continued to decrease to the end of the study. The sharp increase reported by Nixon and associates was not seen by Katkov and co-workers (1982), possibly because values were measured hourly and the early pressure elevations may have returned to baseline levels by the time of the first measurement. On the other hand, the Katkov study demonstrated that mean pulmonary artery pressure gradually increased from 9.6 mm Hg to 11.9 mm Hg over the first 7 hr of bed rest and then gradually decreased to 7.4 mm Hg by the third day of the study. The disparity in right atrial and pulmonary pressures, occurring as it did during the period of bed rest-induced diuresis, remains unexplained. Heart rate, stroke volume, and cardiac output did not change significantly during the $-15°$ head-down study.

Head-down tilt has not been associated with the marked increases in heart size seen with water immersion and correlates with the aforementioned findings of no change or a decrease in right heart pressure (Arborelius *et al.*, 1972; Echt et al., 1974; Lange *et al.*, 1974; Lollgen *et al.*, 1981; Nixon *et al.*, 1979; Polese *et al.*,

TABLE III

Representative Head-Down Bed Rest Studies[a]

Duration (days)	No. of studies	No. of subjects	Body position (degrees from horizontal)	Representative study areas
1–2.5	4	34	−5	Central venous pressure and heart dimensions; urinary sodium excretion
5	2	14	0, −4, −8, −12	Physiological responses to head-down bed rest
7	11	73	−4, −5, −6, −8, −15	Comparison of horizontal versus head-down bed rest; insulin responses; cardiorespiratory and cerebral sensory responses; testing of countermeasures
10–11	2	19	0, −4, −6, −8, −12	Comparison of effects of horizontal versus head-down bed rest; responses to LBNP; exercise as countermeasure
14	2	12	0, −4, −5	Comparison of two modes of bed rest
30	4	56	−2, −4, −6, +6	Physiological responses; electrostimulation as countermeasure
45–49	4	45	−4, −6.5	Bioelectric activity of muscles; physical work capacity; electrostimulation of muscles and exercise as countermeasures
60	1	6	−4.5	Effect of hydrostatic factor and physical fitness on orthostatic tolerance after head-down bed rest
100	1	33	−2, −6, +6	Changes in total body fluid and other physiological responses
182	1	18	−2, −6	Changes in nitrogen metabolism with and without exercise

[a]Additional information on head-down bed rest studies may be found in Greenleaf et al. (1982), Nicogossian et al. (1979), and Sandler (1980).

1980). Dogs subjected to acute −90° head-down tilt showed no increase in heart size when X-rayed (Avasthey and Wood, 1974). Rushmer (1959), in an earlier study on dogs, found the left ventricular dimensions to be maximal in the horizontal body position, as compared with either head up (+30°), or head down (−30°). Explanations for such changes include a shift of the heart, an increase in upper body vascular compliance in adapting to headward fluid shifts, and possible shifts in the hydrostatic indifference point for right heart circulation. Concerning

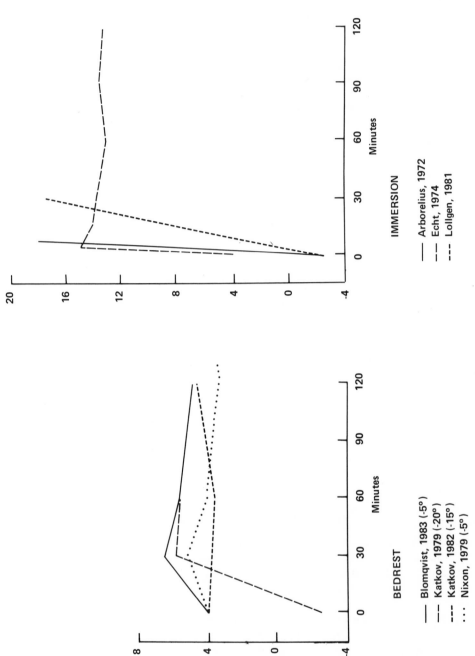

Fig. 5. Changes in venous pressure observed with head-down bed rest and immersion.

the last explanation, Gauer and Thron (1965) reminded investigators that this hydrostatic indifference point designates the level in the hydrostatic columns of the low-pressure side of the circulation where the measured pressure shifts from positive to negative. In the supine position, this point is at midatrium. In the head-up position, it is located just below the diaphragm. In the head-down position, it shifts cephalad to the superior-venacava/atrial junction. A shift in the hydrostatic indifference point helps explain the otherwise puzzling decrease in heart size reported by Jongbloed and Noyons (1933) in rabbits during head-in acceleration studies to a level of -2 G$_z$. Under these conditions, blood was forced from the lower to upper body at twice the force occurring with complete head-down tilt ($-90°$).

Hemodynamic responses during head-down bed rest, in most cases, have paralleled or exceeded those seen with horizontal bed rest, but findings have varied. Soviet investigators have observed no regular increases in resting heart rate during the first month of head-down bed rest but have noted significant decreases thereafter (Panferova, 1976, 1977). Some investigators have found changes in heart rate and cardiac output, whereas others have not (Kovalenko and Gurovsky, 1980; Panferova, 1977). Goldwater and co-workers (1980) in a comparison of 7 days of horizontal and $-6°$ head-down bed rest found similar hemodynamic changes at rest, but a significantly greater fall occurs in post-bed-rest cardiac output during LBNP testing (a 4% decrease at horizontal versus a 39% decrease at $-6°$ head-down, $p < .05$).

Research to date indicates that head-down bed rest is an incisive technique for studying the cardiovascular effects of inactivity and weightlessness. Although many of the changes seen with this technique are seen also with horizontal bed rest, they appear more rapidly and more significantly in the headdown position. More research will be needed with this method, which has been used considerably less than horizontal bed rest, to determine the causes and mechanisms of cardiovascular deconditioning. Future head-down studies will be directed toward the assessing cardiovascular deconditioning occurring with exposure to weightlessness (which the technique closely duplicates), but the basic findings may very well be applicable to conditions on earth resulting from inactivity, immobilization, and aging. (U.S. and Soviet experience with horizontal and head-down bed rest is shown in Tables I and III.)

V. STRESS TEST RESPONSES FOLLOWING INACTIVITY

A. Orthostatic Intolerance

Orthostatic intolerance is a regular feature following bed rest. When the body is placed in an upright position, significant increases in heart rate occur following a few days of bed rest (or 5–6 hr of spaceflight) as compared with responses prior to exposure (Sandler, 1980). In susceptible individuals fainting may occur within a few minutes if they are kept erect. The most commonly used means for testing tolerance have been 70° head-up tilt, passive standing, and LBNP. Head-up tilt at 70°, with the subject passive on a movable table with saddle support, has been used the longest. Prior to bed rest, heart rate usually increases by 15–35 bpm, while 20- to 60-bpm increases are regularly seen after bed rest (Hyatt, 1971; Lancaster and Triebwasser, 1971). Passive standing has been widely used with astronauts and cosmonauts before and after flight. Though easy to apply, it results in a poorer physiological test than tilt, because it promotes muscle contractions in the lower limbs and increases the return of blood pooled in the lower extremities to the central circulation. Such muscle contractions are impossible to control from subject to subject during any given test protocol. Heart rate increases have averaged 35 bpm during 5-min stand tests before bed rest and 60 bpm afterward (Hyatt *et al.*, 1975). Comparative tests have shown that LBNP (from the waist downward) at −40 or −50 mm Hg results in blood pressure and heart rate changes similar to those seen with passive standing and 70° tilt (Wolthuis *et al.*, 1974). The use of LBNP during bed rest studies has shown that 80–90% of subjects (both males and females) can tolerate 15 min of −50 mm Hg before bed rest without fainting. Following bed rest, however, there is a significant increase in heart rate (usually doubling) and a 50% decrease in tolerance time, due to the onset of vasovagal syncope (Greenleaf *et al.*, 1976, 1982; Sandler, 1980). Females undergoing LBNP have shown greater increases in heart rate both before and after bed rest (Montgomery *et al.*, 1977; Sandler and Winter, 1978).

One of the leading explanations for the exaggerated, or deconditioned, response after bed rest has been the possibility of increased venous pooling in dependent body parts during testing, but this hypothesis has not been confirmed. In the vast majority of subjects tested, little change or even a decrease in venous pooling has been found after bed rest or spaceflight (McCally *et al.*, 1971; Sandler, 1980; Sandler and Winter, 1978). Complicating features are the known losses of intravascular volume and lower limb muscle mass with exposure. One of the most careful studies in these regards was conducted by Musgrave and associates (1969), who developed a water plethysmograph to be used with LBNP. Using this approach, leg volumes of bed rested subjects showed smaller or similar increases during LBNP as compared with pre-bed-rest findings (Menninger *et al.*, 1969; Musgrave *et al.*, 1969). Using this same approach, investigators confirmed these findings in a 2-week study of bed rested females who also showed comparable decreases in leg volume during LBNP in 10 out of 12 subjects, despite the regular occurrence of post-bed-rest syncopal reactions (Sandler and Winter, 1978).

An increase in the sympathetic nervous system beta-adrenergic activity with bed rest would also explain the orthostatic losses and would lead to excessive tachycardia and peripheral vascular (arteriolar) vasodilation. The presence of increased sympathetic nervous system tone is supported by the findings of higher resting heart rates in the majority of bed rest subjects. Diastolic pressure also has shown a tendency to rise after bed rest (Polese *et al.*, 1980). The increased sympathetic input is most probably caused by the known decrease in heart volume regularly occurring with bed rest and its triggering of low-pressure cardiopulmonary receptors. Plasma catecholamines at rest are not changed significantly with bed rest (Chobanian *et al.*, 1974; Vernikos-Danellis *et al.*, 1974), but plasma renin activity (PRA) increases have occurred regularly (Chavarri *et al.*, 1977; Goldwater *et al.*, 1980; Melada *et al.*, 1975). Additionally, isoproterenol infusions given after bed rest have produced significantly greater PRA levels, supporting the presence of heightened beta-adrenergic sensitivity (Melada *et al.*, 1975). Findings of Melada and co-workers (1975) and more recently by ourselves (Sandler *et al.*, 1985a) that a beta-adrenergic blockade with propranolol improved, but did not completely remedy, post-bed-rest orthostatic loss also lend support to these conclusions. In contrast, there has been no evidence of change in alpha-adrenergic stimulation, since mean blood pressure responses to graded infusions of norepinephrine and angiotensin II have remained unchanged (Chobanian *et al.*, 1974). Finally, the hemodynamic changes that occur at various levels and stages of orthostatic testing after bed rest closely resemble those that occur in other clinical states such as hemorrhage or shock (Sandler, 1977; Wolthuis *et al.*, 1974), where there is a significant loss of central and circulating blood volume.

Riding on a human centrifuge provides another and a unique opportunity for orthostatic stress testing after deconditioning, since it allows exposures to multiples of the earth's normal gravity field. Interest in physiological reactions to such forces resulted from the redesign of the seating configuration for the space shuttle, which positioned the crew and passengers in a seated position for landing so as to receive a gradual build-up to $+2$ G_z (twice the normal head-to-foot pull of gravity) and to last up to 20 min or more in overall duration. Acceleration testing, ranging from $+1.5$ G_z to $+4$ G_z, has shown expected and progressive post-bed-rest decreases in tolerance times of 50% or more (Goldwater *et al.*, 1977; Sandler, 1980). An unexpected finding was that age and sex did not prove to be deterrents, with subjects up to age 65 years showing similar ability to tolerate exposures up to $+3$ G_z, and older-aged individuals usually showing somewhat better tolerance times, both before and after bed rest (Berry *et al.*, 1980; Goldwater *et al.*, 1981; Sandler *et al.*, 1979). Blood pressure recordings in the older-aged subjects tended to be higher at rest and during accelerations and may have contributed to the maintenance or improvement in tolerance in these cases. The use of G-suits provided consistent and adequate protection in almost all cases, as during orthostatic testing at $+1$ G_z, maintaining adequate cerebral and central circulation by providing physical counterpressure against body fluid displacement to the dependent leg and abdominal regions (Berry *et al.*, 1980; Sandler *et al.*, 1983). Acceleration stress also induced a number of neurohumoral changes, many of which became magnified after bed rest. These included marked increases in plasma adrenocorticotropin (ACTH), plasma fibrinogen, PRA, plasma ADH, and urinary cortisol (Keil and Ellis, 1976; Sandler and Winter, 1978; Vernikos-Danellis *et al.*, 1978). Most changes represent a nonspecific reaction to the stress-testing procedure and serve as a means of documenting the presence and magnitude of the deconditioning process.

B. Exercise Tolerance

Exercise on treadmills or bicycle ergometers has been used extensively to determine what happens to work capacity following bed rest (and spaceflight). After bed rest or spaceflight, exercise tolerance (maximal or submaximal) has decreased significantly, as will be discussed in detail in Chapter 7. Measurements during exercise testing have demonstrated that oxygen uptake, stroke volume, and cardiac output all decrease with bed rest. In contrast, little to no change has been seen in such variables as ventilatory volumes, maximal heart rate, and arteriovenous oxygen differences.

Most investigators have found that maximal oxygen uptake following bed rest decreased by 17–28% (Convertino *et al.*, 1979; Convertino and Sandler, 1982; Hung *et al.*, 1981; Sandler, 1983), although one group found no change (Chase *et al.*, 1966). Five males exposed to 20 days of bed rest showed a decrease of 28% in exercise tolerance when exercising in both the supine and upright positions (Saltin *et al.*, 1968). Since the decrease could not be attributed to impaired venous return during exercise, it was suggested that a primary decrease in myocardial function had been induced by bed rest. More recent studies with middle-aged men, however, have failed to support this conclusion, since a significantly greater decline in $\dot{V}_{O_2 \text{ max}}$ (17% decrease) occurred when the subjects were upright, as compared with 7% when they were supine (Convertino *et al.*, 1982).

The impaired work capacity seen after bed rest (and spaceflight) may very likely stem from physical inactivity, which results in loss of skeletal muscle strength, loss of pumping effectiveness of lower-limb skeletal muscles, and reduced muscle metabolism (Saltin and Rowell, 1980; Scheuer and Tipton, 1977). But these factors have not been thoroughly investigated to date. Although it has been suggested that exercise during immobilization can reduce the dependent cyanosis associated with bed rest (Miller *et al.*, 1964), it has not eliminated the orthostatic intolerance or prevented the loss of acceleration tolerance, although it has lessened the problem somewhat (Greenleaf *et al.*, 1975; Sandler, 1980; Sandler *et al.*, 1985b). It has been difficult to compare the results of the many exercise studies because of differences in study duration and exercise testing methods. Figure 6 illustrates LBNP heart rate and blood pressure changes in one of the seven subjects participating in the repeat bed rest study described in Section IV. Isometric exercise consisted of leg extensions, 30 min twice each day at 21% maximal leg extension force, while isotonic exercise was performed on a supine Collins bicycle ergometer 30 min twice each day at 68% of $\dot{V}_{O_2 \ max}$ (Greenleaf et al., 1975). Analysis of the heart rate changes during LBNP exercise of the magnitude employed failed to counter the deconditioning process. Heart rate levels before and after the second and third bed rest periods tended toward the elevated values seen with the first bed rest exposure. Isometric exercise had no effect during the third exposure. To retard confinement deconditioning between bed rest exposures, the subjects also performed bicycle exercise twice each day (20 min at 50% $\dot{V}_{O_2 \ max}$). Neither this level of exercise nor the exercise performed during bed rest was sufficient to speed the recovery process leading to the next bed rest period, as shown in this figure and Fig. 4.

Recently, evidence has been reviewed by Klein and co-workers (1977), indicating that heavy anaerobic training may be counterproductive in astronauts. Several experiments indicate that athletes decondition at a faster rate than nonathletes, losing a higher proportion of their exercise capacity ($\dot{V}_{O_2 \ max}$) when exposed to higher altitude, water immersion, and bed rest. Furthermore, exercise in-flight or

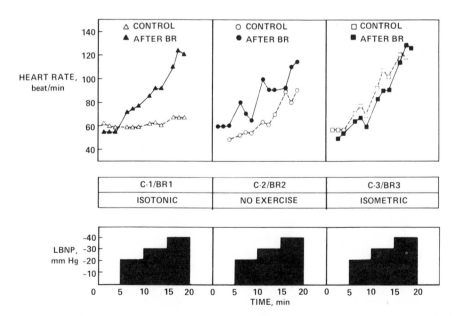

Fig. 6. Heart rate with lower-body negative pressure before and after bed rest, with and without exercise (isotonic and isometric). C-1, C-2, and C-3, pre-bed-rest control periods. BR-1, BR-2, and BR-3, bed rest periods.

during water immersion deconditioning may be more effective in sedentary people or those with a lower $\dot{V}_{O_2\ max}$ (Klein *et al.*, 1977; Stegemann *et al.*, 1975). It is well known that the improvement in exercise capacity after a given amount of training is greater for those starting at a sedentary level (Saltin and Rowell, 1980). In the three-man Skylab flight the two crew members with the highest levels of preflight aerobic capacity and who exercised for up to 2 hr each day had the highest frequency of syncope and presyncope during the in-flight LBNP tests and the greatest loss of orthostatic tolerance and exercise capacity postflight, as compared with the more sedentary third member who performed less in-flight exercise (R. L. Johnson *et al.*, 1977; Klein *et al.*, 1977; Michel *et al.*, 1977).

Several studies have also shown poorer orthostatic tolerance (tilt or LBNP) and poorer acceleration tolerance (head-to-foot, $+G_z$) for athletically conditioned subjects after bed rest (Goldwater *et al.*, 1980, 1981; Korobkov *et al.*, 1968). Similarly, Stegemann and co-workers (1969, 1975) in two separate studies found a much higher incidence of vasodepressor syncope during tilt in athletes, as compared with nonathletes, after 6–8 hr of water immersion. It is interesting that athletes responded as they did despite a much smaller diuresis and plasma volume loss compared with their more sedentary counterparts (Skipka and Schramm, 1982). Based on such findings, the most reasonable explanation for the regularly observed lowered tilt tolerance in athletes both before (Luft *et al.*, 1976, 1980; Mangseth and Bernauer, 1980; Myhre *et al.*, 1976) and after bed rest or water immersion is an alteration in baroreceptor function (Stegemann *et al.*, 1974). However, definitive studies in this regard have not yet been carried out.

VI. CARDIOVASCULAR CHANGES WITH WEIGHTLESSNESS

As of January 1986, the United States had successfully completed 51 manned spaceflight missions and the Soviet Union 59. This has allowed 309 persons access to space (300 men and 9 women), some as many as five time. Ages have ranged from 32 to 57 years. Flight durations have varied from 8 to 10 days. The longest flight on record (237 days) was completed by a Soviet three-man crew on October 4, 1984. Studies during or after flight have shown changes similar or identical to those occurring with bed rest or immersion. After flight, both astronauts and cosmonauts have exhibited postural difficulties, weight loss, significant loss of plasma and blood volume (including red cell mass), and reduced exercise capacity (decreases ranging from 10–50%; Sandler, 1980.) A significant decrease in heart size has also been registered using X-ray or echocardiography (Nicogossian *et al.*, 1977; Henry *et al.*, 1977). Echocardiographic finding immediately after flight in Soviet crews with longer term flight exposures (durations up to 175 days) are shown in Fig. 7 and demonstrate 5–50% losses in left ventricular volumes compared to preflight levels (Yegorov, 1980).

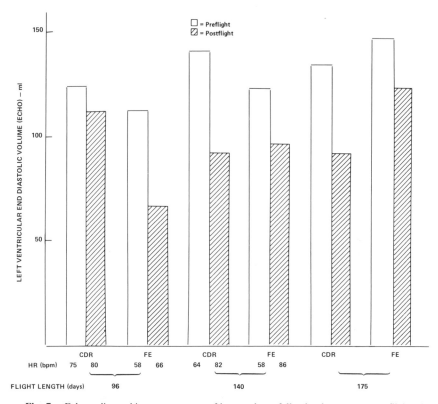

Fig. 7. Echocardiographic measurements of heart volume following long-term spaceflight 96–175 days).

During flight there is a significant headward shift of 1 liter or more of body fluid from the lower half of the body, with attendant symptoms of head fullness, facial puffiness, and neck vein distention, which persist throughout the mission, regardless of length. Such shifts approximate or exceed those seen with bed rest or immersion, as shown in Fig. 1.

Cardiovascular deconditioning appears rapidly during flight. Excess tachycardia, narrowed pulse pressure, and inability to control blood pressure on provocation (LBNP, or exercise tests) were present by at least the second day of flight— the earliest tests conducted to date (Pourcelot *et al.*, 1983). Such changes have persisted throughout flight, but have not been progressive beyond 50–60 days in the 84-day Skylab mission, which is at present the longest U.S. flight of record (R. L. Johnson *et al.*, 1977). Echocardiographic measurements taken during flight aboard American or Soviet missions have shown expected heart volume increases during the first few days; the volume then returned to or below baseline levels as the mission progressed in duration (At'kov and Foyeena, 1985; Bungo *et al.*, 1986; Pourcelot *et al.*, 1983). Findings of postflight cardiovascular deconditioning (exaggerated orthostatic responses, loss of exercise capacity) have occurred despite heavy use of in-flight exercise, LBNP exposure during the last week of flight, and ingestion of fluid (particularly saline) just prior to reentry to replace plasma volume losses. Cardiovacular findings following spaceflight have

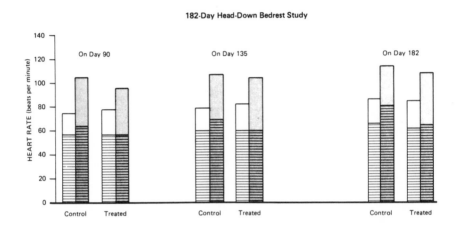

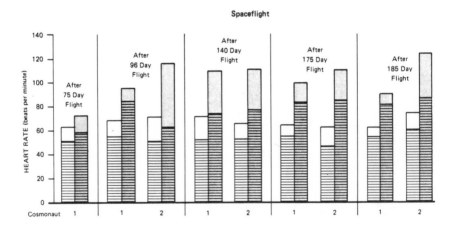

NOTE: Treatment consisted of electrical muscle stimulation.

▤ Horizontal
☐ After 70° upright tilt.
▨ After spaceflight and bedrest.

Fig. 8. Changes in heart rate during 70° tilt before and after prolonged head-down bed rest and spaceflight.

in general been quite similar to those following bed rest. This is shown in Fig. 8, where heart rate response to tilting before and over the course of bed rest (duration up to 182 days) can be seen to be very similar to that occurring before and after flight in Soviet space crews. It should be noted that postflight changes occurred despite the mentioned use of countermeasures for the Soviet flight crews or electrical muscle stimulation in bed rested subjects (Kakurin, 1980).

Both U.S. and Soviet space programs have used animal payloads to both prepare for manned spaceflight and understand physiologic responses (Sandler, 1974). Rats have been most extensively studied, with observations made before and after flights of up to 3 weeks. Specific cardiovascular changes have included electron microscopic findings indicating changes in morphology and function of myocardial cells and altered norepinephrine content for various cardiac chambers, brain stem, and hypothalamus (Kvetnansky et al., 1978, 1982; Kvetnansky and Tigranyan, 1982; Tigranyan et al., 1980). To date, functional implications for these changes have been limited due to the lack of specific in-flight testing procedures, since these animal experiments have been flown aboard unmanned, unattended biosatellites.

VII. MANAGEMENT OF CARDIOVASCULAR EFFECTS OF INACTIVITY

Aerospace medical investigators have spent considerable effort searching for means of counteracting cardiovascular deconditioning. Although these efforts have been directed toward protecting astronauts and cosmonauts, their findings may well apply to conditions of inactivity and immobilization on earth. Exercise has probably received the most attention and is discussed in some detail in Chapter 7. Investigators have hypothesized that strenuous exercise and physical conditioning during inactivity will bring into play the benefits of increased intra-vascular volume and improved skeletal muscle strength and tone. Although supine bicycle exercise during bed rest has tended to lessen post-bed-rest orthostatic intolerance, it has not eliminated it (Birkhead et al., 1964, 1966; Cardus, 1966; Greenleaf et al., 1975; Katkovskiy and Buzulina, 1980; Miller et al., 1964). Furthermore, leg isometric exercise, as already discussed, also failed to impede the deconditioning process (see Figs. 4 and 6). Bungee cord exercises (primarily isometric exercise) did not provide protection against orthostatic intolerance following either bed rest or spaceflight (Berry et al., 1966; Vogt et al., 1967). Heavy exercise regimens (primarily isotonic) during flight with bicycle ergometers or treadmills have created a feeling of well-being in crew members bu have not prevented an altered postflight cardiovascular state, including orthostatic intolerance. Since muscle atrophy is a known and consistent feature of inactivity, further study is needed to determine the optimal exercise regimens

for alleviating such change and to evaluate whether the lack of exercise has an important role in promoting the physiological changes seen with bed rest (or spaceflight). More importantly, careful study will be needed to determine whether its use postflight shortens the readaptation period. Saltin and co-workers (1968) showed that such shortening occurred with exercise in sedentary subjects, but not in those athletically trained.

Deliberate, or induced, venous pooling of the lower extremities has also been tried to avert loss of venous compliance from lack of blood volume displaced to the legs with a horizontal body posture. Periodically inflated cuffs placed on the legs alone, or on both legs and arms, initially produced encouraging results (Graveline, 1962; Vogt and Johnson, 1967), but later work showed that this approach failed to prevent plasma volume losses or to improve tolerance to LBNP and tilt (McCally and Wunder, 1971). When the cuffs were used during spaceflight (Gemini IV and VII), they did not prevent cardiovascular deconditioning. Induced venous pooling using a specially designed suit (Sandler et al., 1983) did improve post-bed-rest losses of exercise capacity, but did not prevent a decrease in LBNP tolerance. Lower-body negative pressure itself has shown some promise when applied repeatedly for 4–6 hr per day during the bed rest period. It has improved plasma volume and orthostatic tolerance (Aleksandrov and Kochetov, 1974; Asyamalov et al., 1973; Stevens et al., 1966). Unfortunately, in most cases, this approach is impractical, because of the long periods of treatment required. Use of LBNP and ingestion of saline just prior to reentry has improved plasma volume and orthostatic tolerance in returning cosmonauts (Gazenko et al., 1981), but has not totally reversed deconditioning. Ingestion of various fluids alone, or the use of steroids (9-fluorohydrocortisone, deoxy-corticosterone, aldosterone), has not been effective in bed rest or water immersion studies (Bohnn et al., 1970; Hyatt, 1971; McCally and Wunder, 1971; Parin et al., 1970; Stevens and Lynch, 1965). Similarly, pitressin administered during water immersion and bed rest prevented diuresis and improved plasma volume but did not prevent orthostatic intolerance (McCally and Wunder, 1971).

A number of pharmacologic agents also have been tested as countermeasures, including central nervous system stimulants (amphetamine, strychnine, and caffeine), androgens (nerobol), and the calcium-blocking agent, isoptin (Bogolyubov et al., 1978; Grigoriev et al., 1976; Parin et al., 1970; Pestov et al., 1969; Trushinskiy et al., 1977). These agents resulted in only partial and incomplete restorative effects on cardiovascular and fluid-electrolyte changes and on muscle degeneration. Atropine has also been tried without success (Chae and Jun, 1979; Murray and Shropshire, 1970). As already mentioned, beta-adrenergic blockade improved tilt tolerance after 5–14 days of bed rest (Melada et al., 1975) and prolonged tolerance to maximal LBNP exposures (5-min steps to -100 mm Hg pressure) (Sandler et al., 1985a). After bed rest systemic vascular resistance was maintained, and even slightly increased, with propranolol, despite a fall in cardiac output, indicating a benefit from beta-2-adrenergic blockade or the release of adrenal steroids or both. Recently, clonidine, an antihypertensive agent that inhibits central nervous system (medullary) sympathetic cardiac accelerator and vasoconstrictor centers has improved orthostatic tolerance in head-down ($-6°$) bed rested subjects (Bonde-Petersen et al., 1981; Guell et al., 1982; Norsk et al., 1981). The use of this agent as a countermeasure after spaceflight is being further pursued by French investigators.

Cooling the mean skin temperature of the lower parts of the body has also been investigated by the use of specially designed, water-cooled garments that reduced the temperature by 10°C (Raven et al., 1980, 1981). This approach has improved LBNP tolerance, increasing resting stroke volume by 11% and by 35% during -50 mm Hg suction. It also has lowered heart rates by 8 bpm, increased blood pressure by 8 mm Hg, and improved muscle tone, as indicated by a significant increase in oxygen uptake ($+84$ ml/min). Although the procedure appears promising, it has not yet been tested in bed rested subjects or during spaceflight.

Soviet investigators have been encouraged by results of studies using electrostimulation of muscles of the lower parts of the body, whether applied alone or in conjunction with physical training or other techniques (Balakhovskiy et al., 1972; Cherepakhin et al., 1977; Georgiyevskiy et al., 1979; Kakurin et al., 1976c, 1980; Yegorov et al., 1970). The approaches used have varied considerably in terms of number of electrodes (10–20), type of portable device, amount of current applied, and the duration of exposure (15 min to several hours a day). The technique was directed toward determining whether bone and muscle deterioration during immobilization could be avoided, but it did not completely prevent cardiovascular deconditioning, as indicated by responses to tilt or LBNP. However, Kakurin and co-workers (1980) noted marked benefits during maximal exercise testing in the 182-day bed rest study. Exercise capacity remained intact in the electrostimulated group but declined by about 40% in the untreated group.

To date, the most effective means of managing inactivity-induced cardiovascular deconditioning appears to be a counterpressure garment (antigravity suit) or leotard. Through physical compression these garments prevent displacement of body fluids to the lower extremities and/or abdomen. The suits have successfully protected young and older-aged subjects from syncope following 2–4 weeks of bed rest (Miller et al., 1964; Sandler et al., 1979). Such garments have been used regularly to prevent fainting in individuals suffering from postural hypotension resulting from autonomic insufficiency (Ibrahim et al., 1975). Both U.S. and Soviet space crews regularly have worn conventional or modified G-suits to protect them from orthostatic intolerance following flight.

Of the many other techniques investigated, such as trampoline jumping, centrifugation, and hypoxia, most are either impractical for use with clinical bed rest or have not shown proven benefits (Bhattacharya et al., 1979; Lamb, 1965;

White *et al.*, 1966). Much research is still needed before an effective means is developed for managing all aspects of cardiovascular deconditioning resulting from inactivity.

VIII. CLINICAL APPLICATIONS

Bed rest studies have overwhelmingly shown that inactivity and immobilization lead to significant changes in physiological function for almost all body systems, with accompanying biochemical and/or histochemical changes as well. The major effects are summarized in Table IV. Many variables show a decrease from initial pre-bed-rest state, others an increase, and still others no change. As repeatedly pointed out in this volume, the negative effects of bed rest result from the loss of hydrostatic pressures within the cardiovascular system, the decrease in metabolism due to inactivity, the loss of skeletal loading, and the psychological stress from isolation and confinement. In many cases, disease can be made worse because of bed rest and in some instances lead to irreversible problems. The associated clinical risks of bed rest have now been covered in a number of books and excellent reviews (Asher, 1947; Browse, 1965; Dock, 1944; Harrison, 1944; Levine, 1944; Levine and Lown, 1951; Powers, 1944; Steinberg, 1980) and have been the primary stimulus for present practices of early ambulation and physical activity after surgery and for certain diseases, such as myocardial infarction. Although clinical complications with bed rest have proved to be valid concerns, quantitation has been made difficult, because data were collected in patients with a wide variety of illnesses, which themselves were the basic cause for the inactivity or placement in bed. Many of the associated illnesses were capable of triggering the observed problems alone, with inactivity contributing, but not being the initiating factor. This became clearer when otherwise healthy individuals were studied, and the incidence of complications was noted to be much smaller and less severe in nature. Clinical findings at the Ames Research Center over the period of 1971–1985 have shown no incidence of thrombosis, embolism, or renal calculi. During this period 167 individuals (119 males and 48 females) were studied, with ages ranging from 19 to 65 years. They were placed in bed for periods ranging from 72 hr to 21 days, with the great majority of subjects (151 individuals) having 7- to 14-day exposures. Findings consisted of earaches, skin rashes, constipation, furuncles, and a few cases of urinary tract infections (females). In a series of longer-term bed rest studies (some lasting 6–8 months) conducted at the U.S. Public Health Service Hospital in San Francisco over a 17-year period (1964–1981), the subjects showed no incidence of major cardiovascular or renal problems. The major clinical problems were earaches and skin infections. In retrospect, it seems clear that the severity of clinical complications are best gauged by the associated, or accompanying, clinical disorders.

TABLE IV

Major Effects of Prolonged Bed Rest on Physiological Function

Decreased	Increased	No change
Orthostatic tolerance (stand, tilt, LBNP)	Syncopal episodes	Resting or exercise arteriovenous oxygen differences
Acceleration tolerance	Resting heart rate	
$\dot{V}_{O_2 max}$	Diastolic blood pressure	Vital capacity
Plasma volume	Max heart rate (stress, exercise)	Maximal voluntary ventilation
Total blood volume	Diuresis	Total lung capacity
Heart and ventricular volumes	Nitrogen loss	Proprioceptive reflexes
Coronary blood flow	Urinary calcium, phosphorus	
Resting cardiac output	Constipation	
Red cell mass, production	Serum fibrinogen	
Sweating threshold	Cholesterol	
Cerebrovascular tone	Low-density lipoproteins	
Pulmonary capillary blood volume	Growth hormone	
Total lung diffusing capacity	ECG changes (ST–T)[a]	
Serum electrolytes	Renal diurnal rhythms	
Hormones (adrenal, ADH)	Deep vein thrombosis	
Blood coagulation	Urinary tract infections	
Balance	Sleep disturbances	
Muscle mass, strength	Psychosocial dissociation	
Bone calcium, density		
Serum proteins		
Insulin sensitivity		
Resistance to infection		
Visual acuity		
Manual coordination		

[a] ST = ST segment; T = T wave depressions.

The present generation is much more conscious of the need for physical activity and sports than any previous, as witnessed by the increase in jogging, health spas, and participation in marathon competitions. Much of this emphasis has been stimulated by the realization that cardiovascular disease is the leading cause of death in the industrialized world, claiming three times as many individuals as does its nearest competitor, cancer. Inactivity has been pointed to as a possible risk or contributing factor (Froehlicher and Oberman, 1972; Kanel *et al.*, 1971, 1976; Morris *et al.*, 1980) and as a complicating factor in individuals past the age of 50, where reduced activity may lead to progressive physical and physiological debility (Harris and Frankel, 1977). Furthermore, response to exercise has become an important and well-established procedure in the diagnosis and evaluation of coronary heart disease (Bruce, 1977; Froelicher, 1983). Chronic use of exercise, which requires a regular habit of aerobic activity at levels greater than that usually performed, results in physical conditioning, or training, as opposed to deconditioning with inactivity. Almost all observed physiological changes with exercise are at the opposite extreme of immobilization and of those shown in Table IV.

Human studies have generally shown significant exercise induced increases in plasma volume, left ventricular mass, and improved capacity for work (Blomqvist and Saltin, 1983; Convertino et al., 1980; Gilbert et al., 1977; Rashkoff, 1976; Scheuer and Tipton, 1977). Results in animals have shown improvement in peripheral blood flow (Blomqvist and Saltin, 1980; Fixler et al., 1976); favorable changes in skeletal muscle mitochondria and enzymes (Ljungqvist and Unge, 1978; Scheuer and Tipton, 1977); increased cardiac mechanical and metabolic performance (Holloszy, 1971; Penpargkul and Scheuer, 1970); age-dependent myocardial hypertrophy (Froehlicher, 1972); and myocardial microcirculatory changes, including increase in coronary size (Burt and Jackson, 1965; Malik, 1977). Exercise, however, has not provided all the benefits hoped for, since there have been mixed results in the development of collateral circulation for the heart and little, if any, effects on established atherosclerotic lesions and risk factors (Carlson, 1967; McAllister et al., 1959; Pedersoli, 1978). Therefore, despite induced physiological changes, a clear-cut role has not yet been demonstrated for exercise training in the treatment and prophylaxis of heart disease, and its use still remains highly controversial, with strong advocates on both sides (Morris et al., 1980; Owen et al., 1980; Thompson et al., 1979; World Health Organization, 1978). Importantly, as already noted, reasonably heavy use of isotonic exercise has not totally prevented induced cardiovascular deconditioning with bed rest or spaceflight; however, its regular use has contributed to an improved state of subject well-being and to the recovery process, particularly when used in nonexercised trained individuals.

One problem in quantitatively evaluating bed-rest-induced deconditioning in normal or ill subjects is that most physiological functions are measured in the supine position, even though humans spend 70% of their waking hours either seated or standing. In the past this situation has occurred, because critical physiological and hemodynamic measurements could only be made in the horizontal position. Now, much of this has been remedied by technological advances, which not only allow subjects to sit upright, but in some cases to be free from direct attachment to the recording system itself. Using biobelt tape recorders, heart rates can be recorded for days, or even weeks, while patients go about their daily pursuits (Arzbaecher, 1978; Stern and Tzivoni, 1974). Similarly, noninvasive measurements of cardiac dimensions (echocardiography and radioisotopes) and intravascular pressure and flow (Swan-Ganz catheters) can be accomplished without restriction of movement in the upright or seated positions (Gilbert et al.,

1977; Hung *et al.*, 1981; Sandler *et al.*, 1977; Swan and Ganz, 1974; Swan *et al.*, 1970). Resultant continuous direct measurements of blood pressure have provided much needed data on the nature of hypertension and drug response (Raferty, 1980). Similar measurements of right heart pressure during several weeks of normal activity and then during a 1-week period of $-15°$ head-down bed rest have clearly indicated that right heart pressures do not show many of the changes that had previously been predicted along theoretical lines (Katkov *et al.*, 1982).

The therapeutic benefits of bed rest are associated with the elimination of hydrostatic gradients, particularly for the lower extremities, postoperative or posttraumatic relief of pain from immobilization of an injured part, and promotion of wound healing. Restriction of motion has also been helpful in the treatment of bone diseases (including fracture and inflammation), tuberculosis and heart disease. Prior to the availability of appropriate chemotherapy, tuberculosis patients were often partially bed rested for several years, and more recently subjects with primary or alcoholic myocardial disease have been treated with strict bed rest for periods of up to 3 years (McDonald *et al.*, 1971, 1972). But, as already discussed, such long periods of bed rest are associated with known definite physiological costs. The use of bed rest for periods longer than 3–5 days, as indicated earlier, leads to a clear-cut manifestation of the deconditioning process. Such knowledge and findings have now led to early ambulation and exercise, which are being increasingly used to reduce postoperative morbidity and for the treatment of myocardial infarction and its rehabilitation process (AHA Committee Report, 1981; Bloch *et al.*, 1974; Hutter *et al.*, 1973; Mayou *et al.*, 1981; McNeer *et al.*, 1975). The goal of such programs has been to ameliorate and/or to reduce the adaptive effects shown in Table IV. DeBusk and Hung (1982), in observing 196 patients undergoing treadmill exercise testing 3 weeks after myocardial infarction, concluded that exercise conditioning proved beneficial for recovering subjects by lowering resting and exercise heart rates, increasing absolute bicycle or treadmill exercise performance, and providing subjective feelings of well-being. Although these responses are indicative of symptomatic and physiological improvement, their use is still controversial; it is not yet known how they may alter prognosis, particularly their effect on the risk of sudden death. As a point of contrast, Sivarajan and co-workers (1981) could not demonstrate a clear-cut benefit for an in-house exercise program in myocardial infarct subjects prior to their hospital release.

Other physiological benefits of an upright body position have been demonstrated, since, as previously discussed, heart size is greatest in the supine position. In the few human studies where heart size and venous pressure have been measured, both have been found to decrease with changes from the supine (Blomqvist and Stone, 1983; Gauer and Thron, 1965; Wilkins *et al.*, 1950), with similar findings in animals. Since cardiac output falls significantly, when the body is upright, cardiac work, which is the product of mean blood pressure and cardiac output, also falls 20–30% (Browse, 1965; Coe, 1954). Such changes were empirically the basis for the use of chair rest, as opposed to strict bed rest, for the treatment of heart disease (Levine and Lown, 1951). Orthostatic unloading by peripheral volume shift may also help explain the relief of symptoms (particularly shortness of breath) during congestive heart failure. In comparison, it is also important to note the original observations made by Abelman and Fareeduddin (1967) and later by Murray and co-workers (1969) of the resistance to orthostatic stress in subjects with congestive heart failure. Maintenance of right heart distention and filling due to intravascular volume overload is protective in these instances.

In permanently inactive individuals, such as paraplegics and quadriplegics, adrenal changes may play an important role in immobilization deconditioning. One study of patients with spinal cord injuries (Kaplan *et al.*, 1966) showed that in 60% of these patients, 17-ketosteroid and 17-hydroxysteroid did not increase with the administration of ACTH. Although healthy individuals exhibit an increase in plasma cortisol concentration with passive tilt, quadriplegic patients did not (Vallbona *et al.*, 1966). The quadriplegic patients, however, excreted more than three times the normal amount of corticoids in the urine. On the basis of these findings, Steinberg (1980) concluded that quadriplegic patients often suffer from chronic stress leading to adrenocortical exhaustion and, consequently, should be given adrenocorticoids to prevent severe hypotension in the upright position. It has also been noted that prolonged immobilization significantly alters circadian, endocrine, and metabolic functions, with the changes persisting for 10 days after reambulation despite the use of exercise (Vernikos-Danellis *et al.*, 1974). (For a more complete discussion of these changes, see Chapters 5 and 6.)

The treatment of all bed rested or paralyzed patients must be given special attention. Simple measures, such as elastic stockings and abdominal binders, may not be effective, particularly in paraplegic and quadriplegic patients. The serious interference with the balance of ventilation and pulmonary circulation caused by immobilization can be offset by turning the patient every hour. Cardiovascular deconditioning may be ameliorated by the use of exercise during the immobilized period. But cogent answers to these problems remain in the future. If we must put certain patients to bed, we hope that current and future research will provide a means of returning them to the ambulatory state without the present physiologically deconditioned state.

REFERENCES

Abelmann, W.H. and Fareeduddin, K. (1967). *Aerosp. Med.* **38**, 60-65.
AHA Committee Report (1981). *Statement on Exercise Circulation* **64**, 1327A.

Aleksandrov, A.N. and Kochetov, A.K. (1974). *Space Biol. and Aerosp. Med.* **8(1)**, 104-105.

Arborelius, M., Jr., Balldin, U.I., Lilja, B., and Lungren, C.E.G. (1972). *Aerosp. Med.* **43**, 592-598.

Arzbaecher, R. (1978). *Med. Instrum.* **12**, 277-281.

Asher, R.J. (1947). *Br. Med. J.* **2**, 967-968.

Asyamalov, B.F., Panchenko, V.S., Pestov, I.D., and Tikhonov, M.A. (1973). *Space Biol. and Med.* **7(6)**, 80-87.

At'kov, O.Y., and Foyeena, G.A. (1985). *In* "Proceedings, 18th Working Group Socialist Countries on Space Biol. and Med." (A.I. Grigoriev, ed.), pp. 8-9. Gagra, U.S.S.R.

Avasthey, P. and Wood, E.H. (1974). *J. Appl. Physiol.* **37**, 166-175.

Balakhovskiy, I.S., Bakhteyeva, V.T., Beleda, R.V., Biryukov, Ye.I., Vinogradova, L.A., Grigor'yev, A.I., Zakharova, S.I., Dlusskaya, I.G., Kiselev, R.K., Kislovakaya, T.A., Kozyrevskaya, G.I., Noskov, V.B., Orlova, T.A., and Sokolova, M.M. (1972). *Space Biol. and Med.* **6(4)**, 110-116.

Berry, C.A., Coons, D.O., Catterson, A.D., and Kelly, C.F. (1966). *In* "Gemini Midprogram Conference," NASA SP-121, pp. 235-263. National Aeronautics and Space Administration, Washington. DC.

Berry, J., Montgomery, L., Goldwater, D., and Sandler, H. (1980). *Aerosp. Med. Assoc. Preprints*, pp. 70-71.

Bhattacharya, A., Knapp, C.F., McCutcheon, E.P., and Evans, J.M. (1979). *J. Appl. Physiol.* **47**, 612-620. Birkhead, N.C., Blizzard, J.J., Daly, J.W., Haupt, C.J., Issekuta, B., Jr., Myers, R.N., and Rodahl, K. (1964). *In* "WADD-AMRL-TDR-64-61," pp. 1-28. Wright-Patterson AFB, Ohio.

Birkhead, N.C., Blizzard, J.J., Issekuta, B., Jr., and Kodahl, K. (1966). *In* "WADD-AMRL-TR-66-6," pp. 1-29. Wright Patterson AFB, Ohio.

Bloch, A., Maeder, J.P., Haissly, J.C., Felix, J., and Blackburn, H. (1974). *Am. J. Cardiol.* **34**, 152-157.

Blomqvist, C.G. and Saltin, B. (1983). *Ann. Rev. Physiol.* **45**, 169-189.

Blomqvist, C.G. and Stone, H.L. (1983). *In* "Handbook of Physiology" Chapter 28, pp. 1025-1063. American Physiological Society, Bethesda, Maryland.

Blomqvist, C.G., Nixon, J.V., Johnson, R.L., Jr., and Mitchell, J.H. (1980). *Acta Astronaut.* **7**, 543-553.

Blomqvist, C.G., Gaffney, F.A., and Nixon, J.Y. (1983). *The Physiologist* **26**, 581-582.

Bogolyubov, V.M., Anashkin, O.D., Trushinskiy, Z.K., Shashkov, V.S., Shatunia, T.P., and Reva, F.V. (1978). *Voyenno-Meditsinskiy Zhurnal* **11**, 64-66.

Bohnn, B.J., Hyatt, K.G., Kamenetsky, L.G., Calder, B.E., and Smith, W.B. (1970). *Aerosp. Med.* **41**, 495-499.

Bonde-Petersen, F., Guell, A., Skagen, K., and Henriksen, C. (1981). *The Physiologist* **24(6)**, S89-S90.

Browse, N.L. (1965). "Physiology and Pathology of Bed Rest," pp. 1-221. Charles C. Thomas, Springfield, Illinois.

Bruce, R.A. (1977). *N. Engl. J. Med.* **296**, 671-677, 1977.

Bungo, M.W., Charles, J.B., Riddle, J., Roesch, J., Wolf, D.A., and Seddon, R. (1986). *J. Aviat. Space Environ. Med.* **57**, 494.

Burkovskaya, T.Ye., Illyukhin, A.V., Lobachik, V.I., and Zhidkov, V.V. (1980). *Space Biol. and Aerosp. Med.* **14(5)**, 75-80.

Burt, J.J. and Jackson, R. (1965). *J. Sports Med. Phys. Fitness* **4**, 203.

Cardus, D. (1966). *Aerosp. Med.* **37**, 993-999.

Carlson, L.A. (1967). *Fed. Proc.* **26**, 1755-1759.

Chae, E. and Jun, S.Y. (1979). *Aerosp. Med. Assoc. Preprints*, pp. 91-92.

Chase, C.A., Grave, C., and Rowell, L.B. (1966). *Aerosp. Med.* **37(12)**, 1232-1238.

Chavarri, M., Ganguly, A., Luetscher, J.A., and Zager, P.G. (1977). *Aviat. Space Environ. Med.* **48**, 633-636.

Cherepakhin, M.A., Kakurin, L.I., Il'ina-Kakuyeva, Ye.I., and Fedorenko, G.T. (1977). *Space Biol. and Aerosp. Med.* **11(2)**, 64-68. Chobanian, A.V., Lille, R.D., Tercyak, A., and Blevins, P. (1974). *Circulation* 49, 551-559.

Coe, W.S. (1954). *Ann. Int. Med.* **40**, 42-48.

Convertino, V.A. and Sandler, H. (1982). *The Physiologist* 25, S159-S160.

Convertino, V.A., Olsen, L., Goldwater, D., and Sandler, H. (1979). *Aerosp. Med. Assoc. Preprints*, pp. 47-48.

Convertino, V.A., Greenleaf, J.E., and Bernauer, E.M. (1980). *J. Appl. Physiol.: Respirat. Environ. Exercise Physiol.* **48**, 657-664.

Convertino, V.A., Hung, J., Goldwater, D., and DeBusk, R.F. (1982). *Circulation* 65, 134-140.

Convertino, V.A., Benjamin, B.A., Keil, L.C. *et al.* (1983). *The Physiologist* **26(4)**, A-60.

deBold, A.J. (1982). *Proc. Soc. Exp. Biol. and Med.* **170**, 133-138.

DeBusk, R.F. and Hung, J. (1982). *Ann. N.Y. Acad. Sci.* **382**, 343-354.

Dewardener, H.E. and Clarkson, E.M. (1982). *Clin. Sci.* **63**, 415-420.

Dock, W. (1944). *JAMA* **125**, 1083-1085.

Echt, M., Dieweling, J., Gauer, O.H., and Lange, L. (1974). *Circ. Res.* **34**, 61-68.

Epstein, M. (1978). *Physiol. Rev.* **58**, 529-581.

Epstein, S.E., Beiser, G.S., Stamfer, M., and Braunwald, E. (1968a). *J. Clin. Invest.* **47**, 139-152.

Epstein, S.E., Stampfer, M., and Beiser, C.D. (1968b). *Circulation* 37, 524-533.

Fixler, D.E., Atkins, J.M., Mitchell, J.H., and Horwitz, L.D. (1976). *Am. J. Physiol.* **231**, 1515.

Froehlicher, V.F. (1972). *Am. Heart J.* **84**, 496-506.

Froehlicher, V.F. (1983). "Exercise Testing and Training," pp. 1-348. Year Book Publishing Co., Chicago, Illinois.

Froehlicher, V.F., and Oberman, A. (1972). *Prog. Cardiovasc. Dis.* **15**, 41-65.

Gauer,O.H. and Henry, J.P. (1976). *In* "International Review Physiology, Cardiovascular Physiology" (A.C. Guyton and A.W. Cowley, eds.), Vol. 9, Sect. II, pp. 145-190. University Park Press, Baltimore, MD.

Gauer, O.H. and Thron, H.L. (1965). *In* "Handbook of Physiology: Circulation" (W.F. Hamilton, ed.), Vol. III, Sect. 2, pp. 2409-2439. American Physiological Society, Washington, DC.

Gauer, O.H., Henry, J.P., and Behn, C. (1970). *Ann. Rev. Physiol.* **32**, 547-595.

Gazenko, O.G., Genin, A.M., and Yegorov, A.D. (1981). *Acta Astronautica* **8**, 907-917.

Genin, A.M. and Kakurin, L.I. (1972). *Space Biol. and Med.* **6(4)**, 42-44.

Genin, A.M., Sorokin, P.A., Gurvich, G.I., Dzhamgarov, T.T., Panov, A.G., Ivanov, I.I., and Pestov, I.D. (1969). *In* "Problems of Space Biology" (A.M. Genin and P.A. Sorokin, eds.), Vol. 13, pp. 256-262. Nauka Press, Moscow.

Georgiyevskiy, V.A., Il'Inskaya, Ye.A., Mayeyev, V.I., Mikhaylov, V.M., and Pervushin, V.I. (1979). *Space Biol. and Aerosp. Med.* **13(6)**, 52-57.

Gilbert, C.A., Nutter, D.O., Felner, J.M., Perkins, J.V., Heymsfield, S.B., and Schlant, R.S. (1977). *Am. J. Cardiol.* **40**, 528-533.

Gilmore, J.P. (1983). *In* "Handbook of Physiology: The Cardiovascular System" (J.T. Shepherd and F.M. Abboud, eds.), pp. 885-915. American Physiological Society, Bethesda, Maryland.

Goetz, K.L., Bond, G.C., and Bloxham,D.D. (1975). *Physiol. Rev.* **55**, 157-205.

Goldwater, D., Sandler, H., Rostiano, S., and McCutcheon, E.P. (1977). *Aerosp. Med. Assoc. Preprints*, pp. 204-241.

Goldwater, D., Melada, M., Polese, A., Keil, L., and Luetscher, J.A. (1980). *Circulation* **62**, III-287.

Goldwater, D.J., Convertino, V., and Sandler, H. (1981). *Aerosp. Med. Assoc. Preprints*, pp. 179-180.

Graveline, D.E. (1962). *Aerosp. Med.* **33**, 297-302.

Greenleaf, J.E. (1983). *In* "Space Physiology" (J. Garcia, M. Guerin, and C. Laverlochere, eds.), pp. 335-348. Toulouse Cepaudes Editions, Toulouse, France.

Greenleaf, J.E., van Beaumont, W., Bernauer, E.M., Haines, R.F., Sandler, H., Staley, R.W., Young, H.L., and Yusken, J.W. (1975). *Aviat. Space Environ. Med.* **46**, 671-678.

Greenleaf, J.E., Greenleaf, C.J., Van Derveer, D., and Dorchak, K. (1976). 'Adaptation to Prolonged Bedrest in Man: A Compendium of Research", pp. 1-183. NASA TM-X-3307, National Aeronautics and Space Administration, Washington, DC.

Greenleaf, J.E., Silverstein, L., Bliss, J., Langenheim, V., Rossow, H., and Chao, C. (1982). "Physiological Responses to Prolonged Bed Rest: A Compendium of Research (1974-1980)." NASA TM-81324, National Aeronautics and Space Administration, Washington, DC.

Grigoriev, A.I., Pak, Z.P., Koloskova, Yu.S., Kozyrevskaya, G.I., Korotayev, M.M., and Bezumova, Yu.Ye. (1976). *Space Biol. and Aerosp. Med.* **10(4)**, 83-89.

Guell, A., Gharib, Cl., Gauquelin, G., Montastruc, P., and Bes. A. (1982). *The Physiologist* **25(4)**, S69-S70.

Hagan, R.D., Diaz, F.J., and Horvath, S. (1978). *J. Appl. Physiol.* **45(3)**, 414-417.

Harris, R. and Frankel, L.J. (eds.), (1977). "Guide to Fitness After Fifty," pp. 1-356. Plenum Press, New York and London.

Harrison, T.R. (1944). *JAMA* **125**, 1075-1077.

Henry, W.L., Epstein, S.E., Griffith, J.M., Goldstein, R.F., and Redwood, D.R. (1977). *In* "Biomedical Results of Skylab," NASA SP-377, pp. 366-371. National Aeronautics and Space Administration, Washington, DC.

Holloszy, J.O. (1971). *In* "Coronary Heart Disease and Fitness," (O.A. Larsen and R.O. Malmborg, eds.), pp. 147-151. University Park Press, Baltimore, Maryland.

Hung, J., Goldwater, D., Convertino, V., McKillop, J., Goris, M., and DeBusk, R. (1981). *Am. J. Cardiol.* **47**, 477.

Hutter, A.M., Jr., Sidel, V.W., Shine, K.L., amd DeSanctis, R.W. (1973). *N. Engl. J. Med.* **288**, 1141-1144.

Hyatt, K.W. (1971). *In* "Hypogravic and Hypodynamic Environments", (R.H. Murray and N. McCally, eds.), pp. 187-209. NASA-SP-269, National Aeronautics and Space Adminisration, Washington, DC.

Hyatt, K.H., Sullivan, R.W., Spears, W.R., and Vetter, W.R. (1973). U.S. Public Health Service Hospital, San Francisco, NASA Contract Report T-81035, pp. 1-75.

Hyatt, K.H., Jacobson, L.B., and Schneider, V.S. (1975). *Aviat. Space Environ. Med.* **46**, 801-806.

Ibrahim, M., Tarazi, R.C., and Dustan, H.P. (1975). *Am. Heart J.* **90**, 513-520.

Johnson, P.C., Driscoll, R.E., and LeBlanc, A.D. (1977). *In* "Biomedical Results from Skylab," pp. 235-241. NASA SP-377, National Aeronautics and Space Administration, Washington, DC.

Johnson, R.L., Hoffler, G.W., Nicogossian, A.E., Bergman, S.A., Jr., and Jackson, M.M. (1977). *In* "Biomedical Results of Skylab," NASA SP-377, pp. 284-312. National Aeronautics and Space Administration, Washington, DC.

Jongbloed, J. and Noyons, A.K. (1933). *Pfluegers Arch. Ges. Physiol.* **233**, 67-97.

Kakurin, L.I., Kuzman, M.P., Matsnev, E.I., and Mikhailov, V.M. (1976a). *In* 'Life Sciences and Space Research", (P.N.A. Sneath, ed.), Vol. XIV, pp. 101-108. Academie Verlay, Berlin.

Kakurin, L.I., Lobachik, V.I., Mikhaylov, M., and Senkevich, Y.A. (1976b). *Aviat. Space Environ. Med.* **47**, 1084-1086.

Kakurin, L.I., Yegorov, B.B., Il'ina, Ye.I., and Cherepakhin, M.A. (1976c). *In* "International Symposium on Basic Environmental Problems of Man in Space" (A. Graybiel, ed.), pp. 241-247. Pergamon Press, Oxford.

Kakurin, L.I., Grigor'yev, A.I., Mikhaylov, V.M., and Tishler, V.A. (1980). *In* "Proceedings 11th US/USSR Joint Working Group Meeting on Space Biology and Medicine," Moscow. NASA-TM-76465, pp. 1-57, National Aeronautics and Space Administration, Washington, DC.

Kanel, W.B., McGee, D., and Gordon, T. (1976). *Am. J. Cardiol.* **38**, 46-51.

Kaplan, L., Powell, B.R., Grynbaum, B.B., and Rusk, H.S. (1966). 'Comprehensive Follow-up Study of Spinal Cord Dysfunction and Its Resultant Disabilities", pp. 1-184, Institute of Rehabilitation Medicine, New York.

Katkov, V.Ye., Chestukin, V.V., Lapteve, R.I., Yakoleva, V.A., Mikhaylov, V.M., Zybin, D.K., and Utkin, V.N. (1979). *Aviat. Space Environ. Med.* **50**, 147-153.

Katkov, V.Ye., Chestukhin, V.V., Nikolayenko, E.M., Grozdev, S.V., Rumyantsev, V.V., Guseynova, T.M., and Yegorova, I.A. (1982). *Space Biol. and Aerosp. Med.* **16(5)**, 64-72.

Katkovskiy, B.S., and Buzulina, V.P. (1980). *Space Biol. and Aerosp. Med.* **14(3)**, 86-87.

Keil, L.C., and Ellis, S. (1976). *J. Appl. Physiol.* **40**, 911-914.

Kimsey, S.L. (1977). *In* "Biomedical Results from Skylab" (R.S. Johnston and L.F. Dietlein, eds.), pp. 249-284. NASA SP-377, National Aeronautics and Space Administration, Washington, DC.

Klein, K.R., Wegman, H.M., and Kuklinski, P. (1977). *Aviat. Space Environ. Med.* **48**, 215-222.

Korobkov, A.V., Ioffee, I.A., Abriksova, M.A., and Stoyda, Yu.M. (1968). *Space Biol. and Med.* **2(3)**, 48-57.

Korolev, B.A. (1969). *Space Biol. and Med.* **2(5)**, 96-101.

Kovalenko, Ye.A. and Gurovskiy, N.N. (1980). *In* "Hypokinesia" (Russian), Chapter 3, pp. 107-208. Meditsina, Moscow.

Krasnykh, I.G. (1973). *Voyenno-Meditsinskiy Zhurnal* **12**, 54-56.

Krasnykh, I.G. (1974). *Space Biol. and Aerosp. Med.* **8(1)**, 98-103.

Krasnykh, I.G. (1979). *Space Biol. and Aerosp. Med.* **13(5)**, 40-45.

Kvetnansky, R., and Tigranyan, R.A. (1982). *Space Biol. and Med.* **16(4)**, 139-142.

Kvetnansky, R., Kopin, I.J., and Saavedra, J.M. (1978). *Brain Res.* **115**, 387-390.

Kvetnansky, R., Culman, J., Serova, L.V., Tigranyan, R.A., Torda, T., and Macho, L. (1982). *In* "Proceedings 33rd International Astronautical Congress", pp. 94-100, Paris.

Lamb, L.E. (1965). *Aerosp. Med.* **36**, 97-100.

Lancaster, M.D. and Triebwasser, J.H. (1971). *In* "Hypogravic and Hypodynamic Environments" (R.H. Murray and M. McCally, eds.), NASA SP-269, pp. 225-248. National Aeronautics and Space Administration, Washington, DC.

Lange, L., Lange, S., Echt, M., and Gauer, O.H. (1974). *Pfluegers Arch.* **353**, 219-226.

Levine, S.A. (1944). *JAMA* **126**, 80-84.

Levine, S.A. and Lown, B. (1951). *Trans. Assoc. Am. Physic.* **64**, 316-327.

Ljungqvist, A. and Unge, G. (1978). *J. Appl. Physiol.* **43**, 306-307.

Lollgen, H., Just, H., Wollschlages, H., and Kersting, F. (1981). *Z. Kardiol.* **70(5)**, 425-428.

Luft, U.C., Myhre, L.G., Loeppky, J.A., and Venters, M.D. (1976). *In* "NASA Contr. Report NAS9-14472," pp. 2-60. Johnson Space Center, Houston, Texas.

Luft, U.C., Loeppky, J.A., Venters, M.D., and Kobayashi, Y. (1980). "Tolerance of LBNP in Endurance Runners, Weightlifters, Swimmers, and Non-athletes." NASA Contr. Report NAS9-15483, pp. 1-35. Johnson Space Center, Houston, Texas.

Malik, A.B. (1977). *Cardiovasc. Med.* **2**, 1137-1141.

Mangseth, G.R. and Bernauer, E.M. (1980). *Med. and Sci. in Sports and Exercise* **12**, 140.

Mayou, R. Macmahon, D., Sleight, P., and Florencio, M.J. (1981). *Lancet* **2**, 1399.

McAllister, F.F., Bertsch, R., Jacobson, J., and D'Alessio, G. (1959). *Arch. Surg.* **80**, 54-60.

McCally, M. and Wunder, C.C. (1971). *In* "Hypogravic and Hypodynamic Environments", (R.H. Murray and M. McCally, eds.), NASA SP-269, pp. 323-344. National Aeronautics and Space

Administration, Washington, DC.

McCally, M., Kazarin, L.E., and von Gierke, H.E. (1971). *In* "Proceedings, 21st International Astronautical Congress," pp. 264-282.

McDonald, C.D., Burch, G.E., and Walsh, J.J. (1971). *Ann. Intern. Med.* **74**, 681-688.

McDonald, C.D., Burch, G.E., and Walsh J.J. (1972). *Am. J. Med.* **52**, 41-50.

McNeer, J.F., Wallace, A.G., Wagner, G.S., Starmer, C.F., and Rosati, R.A. (1975). *Circulation* **51**, 410-413.

Melada, G.A., Goldman, R.H., Luetscher, J.A., and Zager, P.G. (1975). *Aviat. Space Environ. Med.* **46**, 1049-1055.

Menninger, R.P., Mains, R.C., Zechman, F.W., and Piemme, T.A. (1969). *Aerosp. Med.* **40**, 1323-1326.

Michel, E.L., Rummel, J.A., Sawin, C.G., Buderer, M.C., and Lem, J.D. (1977). *In* "Biomedical Results From Skylab," pp. 284-312. NASA SP-377, National Aeronautics and Space Administration, Washington, DC.

Miller, P.B., Hartman, B.O., Johnson, R.L., and Lamb, L.E. (1964). *Aerosp. Med.* **35**, 931-939.

Montgomery, L.D., Kirk, P.J., Payne, R.A., Gerber, R.L., Newton, S.D., and Williams, B.A. (1977). *Aviat. Space Environ. Med.* **48**, 138-145.

Morris, J.N., Pollard, R., Everitt, M.G., Chave, S.P., and Semmence, A.M. (1980). *Lancet* **2**, 1207-1210.

Morse, B.S. (1967). *In* "Lectures in Aerospace Medicine," (6th Series), pp. 240-254. AD-665-107, School of Aerospace Medicine, Brooks AFB, Texas.

Murray, R.H. and Shropshire, S. (1970). *Aerosp. Med.* **41**, 717-722.

Murray, R.H., Thompson, L.J., Bowers, J.A., Steinmetz, E.F., and Albright, C.C. (1969). *Circulation* **39**, 55-63.

Musgrave, F.S., Zechman, F.W., and Main, R.S. (1969). *Aerosp. Med.* **40**, 602-606.

Myhre, L.G., Luft, U.C., and Venkers, M.D. (1976). *Med. Sci. Sports* **8**, 53-54.

Nelson, C.V., Rand, P.W., Angelakos, E.T., and Hugenholtz, P.G. (1972). *Circ. Res.* **31**, 95-104.

Nicogossian, A.E., Hoffler, G.W., Johnson, R.L., and Gowen, R.J. (1977). *In* "Biomedical Results from Skylab," NASA SP-377, pp. 400-407. National Aeronautics and Space Administration, Washington, DC.

Nicogossian, A.E., and Sandler, H., Whyte, A.A., Leach, C.S., and Rambaut, P.C. (1979). Natl. Aeronaut. Space Admin. Biotechnol. Inc. and Office of Life Sciences, pp. 1-370.

Nixon, J.V., Murray, R.H., Bryant, C., Johnson, R.L., Mitchell, J.H., Holland, O.B., Gomez-Sanchez, C., Vergne-Marini, P., and Blomqvist, C.G. (1979). *J. Appl. Physiol.* **46**, 541-548.

Norsk, P., Bonde-Peterson, F., and Warberg, J. (1981). *The Physiologist* **24(6)**, S91-S92.

Owen, C.A., Pnor, B.W., Beard, E.F., and Jackson, A.S. (1980). *J. Occupat. Med.* **22**, 235-240.

Padfield, P.L., and Morton, J.J. (1983). "Handbook of Hypertension: vol. 1. Clinical Aspects of Essential Hypertension" (J.L.S. Robertson, ed.), pp. 348-364. Elsevier, Amsterdam.

Panferova, N.Ye. (1976). *Space Biol. and Aerosp. Med.* **10(6)**, 18-25.

Panferova, N.Ye. (1977). "Hypodynamia and the Cardiovascular System" (Russian), pp. 1-336. Nauka Press, Moscow. (In English NASA TM-76291.)

Parin, V.V., Krupina, T.N., Mikhaylovskiy, G.P., and Tizul, A.Ya. (1970). *Space Biol. and Med.* **4(5)**, 91-98.

Pedersoli, W.M. (1978). *Cur. Therap. Res.* **23**, 464.

Penpargkul, S., and Scheuer, J. (1970). *J. Clin. Invest.* **49**, 1859.

Pestov, I.D., Tishchenko, M.I., Korolev, B.S., Asyamolov, B.F., Simonenko, V.V., and Baykov, A.Ye. (1969). *Prob. Space Biol.* **13**, 238-247.

Polese, A., Goldwater, D., London, L., Yuster, D., and Sandler, H. (1980). *Aerosp. Med. Assoc. Preprints*, pp. 24-25.

Pourcelot, L., Pottier, J.M., Patat, F., Arbielle, P.H., Kotoovskaya, I., Guuenin, A., Savilov, A., Bistrov, U., Golovkina, D., Bost, R., Simon, P., Guell, A., and Garib, C. (1983). *In* "Proc. 34th Congress International Astronautical Federation," Vol. 2, pp. 92-105.

Powers, J.H. (1944). *JAMA* **125**, 1079-1083.

Raferty, E.B. (1980). *In* "The Methodology of Blood Pressure Recording in the Cardiovascular System" (R.G. Shanks, ed.), pp. 20-28. McMillan Press, London.

Rashkoff, W.J. (1976). *JAMA* **236**, 158-162.

Raven, P.B., Saito, M., Gaffney, F.A., Schutte, J., and Blomqvist, C.G. (1980). *Aviat. Space Environ. Med.* **51**, 497-503.

Raven, P.B., Papa, G., Taylor, W.F., Gaffney, F.A., and Blomqvist, C.G. (1981). *Aviat. Space Environ. Med.* **52**, 387-391.

Rushmer, R.F. (1959). *Circulation* **20**, 897-905.

Saltin, B. and Rowell, L. (1980). *Fed. Proc.* **39**, 1506-1513.

Saltin, B., Blomqvist, G., Mitchell, J.H., Johnson, R.L., Jr., Wildenthal, K., and Chapman, C.B. (1968). *Circulation* **38**, S1-S78.

Sandler, H. (1974). *In* "The Use of Nonhuman Primates in Space" (R. Simmonds and G.H. Bourne, eds.), NASA Conf. Pub. 005, pp. 3-21. National Aeronautics and Space Administration, Washington, DC.

Sandler, H. (1977). *In* "Progess in Cardiology" (P.N. Yu and J.F. Goodwin, eds.), pp. 227-270. Lea and Febiger, Philadelphia, Pennsylvania.

Sandler, H. (1980). *In* "Hearts and Heart-Like Organs" (G.H. Bourne, ed.), Vol. 2, pp. 435-524. Academic Press, New York.

Sandler, H. (1983). *In* "SAE Technical Paper Series, No. 831132", pp. 1-8, Society of Automotive Engineers, Inc.

Sandler, H. and Winter, D.L. (1978). "Physiological Responses of Women to Simulated Weightlessness: A Review of the Significant Findings of the First Female Bed Rest Study," NASA SP-430, pp. 1-87. National Aeronautics and Space Administration, Washington, DC.

Sandler, H., Popp, R., and McCutcheon, E.P. (1977). *Aerosp. Med. Assoc. Preprints*, pp. 242-243.

Sandler, H., Goldwater, D., Rositano, S., Sawin, C., and Booher, C. (1979). *Aerosp. Med. Assoc. Preprints*, pp. 43-44.

Sandler, H., Webb, P., Annis, J.F., Pace, N., Grunbaum, B.W., Dolkas, D., and Newsom, B. (1983). *Aviat. Space Environ. Med.* **54(3)**, 191-201.

Sandler, H., Goldwater, D.J., Popp, R.L., Spaccavento, L., and Harrison. D. (1985a). *Am. J. Cardiol.* **55** 114D-120D.

Sandler, H., Popp, R.L., and Harrison, D.C. (1985b). *Aviat. Space Environ. Med.* **56**, 489.

Scheuer, J. and Tipton. C.M. (1977). *Ann. Rev. Physiol.* **39**,221-251.

Sivarajan, E.S., Bruce, R.A., Almes, M.J., Green, B., Belanger, L., Lindskog, B.D., Newton, K.M., and Mansfield, L.W. (1981). *N. Eng. J. Med.* **305**, 357-362.

Skipka, W. and Schramm, U. (1982). *In* "Zero-G Simulation for Ground-Based Studies in Human Physiology, with Emphasis on the Cardiovascular and Body Fluid Systems," pp. 29-38. ESA SP-180, European Space Agency, Paris.

Stegemann, J., Framing, H.D., and Schiefeling, M. (1969). *Pfluegers Arch.* **312**, 129-138.

Stegemann, J., Busert, J.A., and Brock, D. (1974). *Aerosp. Med.* **45**, 45-48.

Stegemann, J., Framing, H.D., and Schiefeling, M. (1975). *Aviat. Space Environ. Med.* **46**, 26-29.

Steinberg, F.U. (1980). "The Immobilized Patient: Functional Pathology and Management," pp. 1-156. Plenum Medical Book Company, New York and London.

Stern, S. and Tzivoni, D. (1974). *Brit. Heart J.* **36**, 481-486.

Stevens, P.M. and Lynch, T.N. (1965). *Aerosp. Med.* **36**, 1151-1156.

Stevens, P.M., Miller, P.B., Gilbert, C.A., Lynch, T.N., Johnson, R.L., and Lamb, L.E. (1966). *Aerosp. Med.* **37**, 357-367.

Swan, H.J.C. and Ganz, W. (1974). *Am. J. Cardiol.* **34**, 119.

Swan, H.J.C., Ganz, W., Forrester, J., Marins, A., Diamond, G., and Chonette, D. (1970). *N. Engl. J. Med.* **283**, 447-459.

Taylor, H.L., Henschel, A., Brozek, J., and Keys, A. (1949). *J. Appl. Physiol.* **2**, 223.

Thibault, G., Garcia, R., Cantin, N., and Genest, J. (1983). *Hypertension (Suppl. 1)*, I-75 to I-80.

Thompson, P.D., Stern, M.P., William, P., Duncan, K., Haskell, W.L., and Wood, P.D. (1979). *JAMA* **242**, 1265-1267.

Tigranyan, R.A., Macho, L., Kvetnansky, R., Nemeth, S., and Kalita, N.F. (1980). *The Physiologist* **23**, S45-46.

Tkachev, V.V. and Kul'kov, Ye. (1975). *Space Biol. and Aerosp. Med.* **9(1)**, 135-142.

Trushinskiy, Z.K., Bogolyubov, V.M., Anashkin, O.D., Shashkov, V.S., Timofeyev, M.F., Sverdlik, Yu.N., and Reva, F.V. (1977). *Kardiologiya* **12**, 90-94.

Turbasov, V.D. (1980). *Space Biol. and Aerosp. Med.* **14(5)**, 81-87.

Vallbona, C., Lipscomb, H.S., and Carter, R.E. (1966). *Arch. Phys. Med. Rehab.* **47**, 412-421.

Vernikos-Danellis, J., Winget, C.M., Leach, C.S., and Rambaut, P.C. (1974). 'Circadian, Endocrine, and Metabolic Effects of Prolonged Bedrest: Two 56-Day Bedrest Studies," NASA TMX-3051, pp. 1-45. National Aeronautics and Space Administration, Washington, DC.

Vernikos-Danellis, J., Dallman, M.F., Forsham, P., Goodwin, A.L., and Leach, C.S. (1978). *Aviat. Space Environ. Med.* **49**, 886-889.

Vetter, W.R., Sullivan, R.W., and Hyatt, K.H. (1971). *Aerosp. Med. Assoc. Preprints*, 56-57.

Vogt, F.B. and Johnson, P.B. (1967). *Aerosp. Med.* **38**, 702-707.

Vogt, F.B., Mack, P.B., and Johnson, P.C. (1967). *Aerosp. Med.* **38**, 1134-1137.

White, P.D., Nyberg, J.W., Finney, L.M., and White, W.J. (1966). "Influence of Periodic Centrifugation on Cardiovascular Functions of Men During Bed Rest," pp. 1-54. NASA CR-65422, Douglas Aircraft Co., Inc., Santa Monica, California.

Wilkins, R.W., Bradley, S.E., and Friedland, C.K. (1950). *J. Clin. Invest.* 29, 940-949.

Wolthuis, R.A., Bergman, S.A., and Nicogossian, A.R. (1974). *Physiol. Rev.* **34**, 566-595.

World Health Organization (1978). "Habitual Physical Activity and Health." European Series No. 6, WHO, Copenhagen.

Yegorov, A.D. (1980). "Proc. 11th US/USSR Joint Conf. Space Biol. and Med.," NASA TM-76450, pp. 1-225. National Aeronautics and Space Administration, Washington, DC.

Yegorov, B.B., Georgiyevskiy, V.S., Mikhaylov, V.M., Kil, V.I., Semeniutin, I.P., Kaxmirov, E.K., Davidenko, Yu.V., and Fat'ianova, L.I. (1970). *Space Biol. and Med.* **3(6)**, 96-101.

Ziegler, M.S., Lake, C.R., and Kopin, I.J. (1977). *New Engl. J. Med.* **296**, 293-297.

3

Effects of Inactivity on Bone and Calcium Metabolism

SARA B. ARNAUD,* VICTOR S. SCHNEIDER,† AND EMILY MOREY-HOLTON*

*Biomedical Research Division
National Aeronautics and Space Administration
Ames Research Center
Moffett Field, California 94035

†University of Texas Health Science Center
Houston, Texas 77025

Bone is a dynamic tissue that functions not only as mechanical support, but also as a major component of the metabolic and endocrine systems that maintain mineral homeostasis. No situation illustrates the interdependence of these two major functions of bone more clearly than immobilization. In 1941, Albright and co-workers reported the occurrence of severe hypercalcemia in a 14-year-old boy immobilized following a fracture. This case eventually alerted physicians to the disturbances in calcium metabolism during bed rest. Yet, in spite of advances in our knowledge of both mineral and skeletal metabolism, which enable us to predict and to treat immobilization hypercalcemia (Schneider and Sherwood, 1974), the precise reason for the increase in blood calcium and disuse osteoporosis still remains elusive. Signals, generated by biomechanical activity on bone, in some way modulate or interact with bone and calcium metabolism but have not yet been characterized. Nor do we understand the biochemical events in bone that lead to mineralization, growth, and the maintenance of its structure during activity and mechanical loading. Exchange of ideas and data by scientists in separate disciplines directed at the study of biomechanical, cellular, orthopedic and orthodontic aspects of bone function is increasing (Cowin et al., 1984) and will lead to the collaborative studies needed to understand how bone tissue functions during different activities in various environments.

The studies to be discussed in this chapter summarize our current knowledge on the effects of bed rest and/or immobilization on bone and calcium metabo-

49

lism. They do not represent an all-inclusive listing of past research but do include landmark studies that have contributed to our current concepts of the effect of bed rest on skeletal and mineral metabolism. Our review suggests that bone, as an organ system, responds or adapts to immobilization from a number of causes in a similar way. We have, therefore, included observations acquired both in healthy individuals who have volunteered for bed rest studies and in patients with spinal cord injury. Although paraplegia and bed rest are clearly different clinical entities, studies in paralyzed patients have contributed significantly to our knowledge and understanding of bone atrophy. We speculate that the effects of spinal cord injuries represent the extreme end of a spectrum of immobilization osteoporosis, while healthy volunteers during bed rest and astronauts in weightlessness are at the other, less severe, end of the spectrum. Single-limb immobilization might fall somewhere between these two extremes of severity.

Following an injury that leads to inactivity, acute and transitional changes in both skeletal metabolism and calcium homeostasis occur, leading to a chronic reduced metabolism of bone. Metabolic and morphological characteristics of each stage are different, and the magnitude and duration of each phase or stage of adaptation may vary among individuals. Figure 1 presents a hypothetical scheme, drawn largely from the work of Minaire and associates (1974), on disuse osteoporosis following spinal cord injury. It describes what we believe to be the general time course of events in bone and calcium metabolism which precedes disuse osteoporosis and finally leads to a more chronic phase in which the metabolism of a reduced bone mass is stabilized. During the longest bed rest study ever conducted (9 months) and later confirmed by 5-month studies, the patterns of calciuria and bone density loss as a function of time were similar to, though less severe than, those in the paralegic (Donaldson et al., 1970; Schneider and McDonald, 1984). A bed rested or immobilized patient presents a challenge to the clinician over the course of confinement. Although management of the problems of hypercalciuria and nephrolithiasis are early acute problems requiring attention, it is not known how treatment during the acute phase will alter the adaptation to inactivity or reduced bone mass which invariably follows.

I. CURRENT CONCEPTS OF SKELETAL METABOLISM

Bone matrix consists of collagen (95%), protein polysaccharides (4.4%, primarily chondroitin sulfate A), noncollagenous proteins (0.5%), and lipids (less than 0.1%). The mineral phase of bone is composed mainly of hydroxyapatite-like crystals $[Ca_{10}(PO_4)_6(OH)_2]$ and amorphous calcium phosphate (Quelch et al., 1983). The long axis of hydroxyapatite crystals parallels collagen fibers. The hydration shell that surrounds hydroxyapatite crystals interfaces with the crystal and the extracellular fluid of bone which is unique in its high content of potassium (Neuman, 1982). The synthesis of noncollagenous bone proteins of

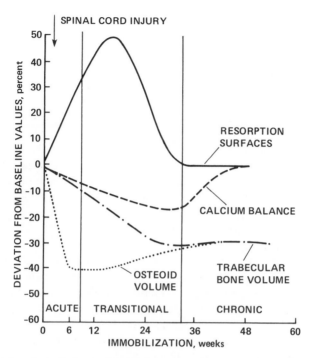

Fig. 1. A hypothetical scheme of the acute transitional and chronic phases of the response of bone histology and calcium balance in patients with acute disuse osteoporosis secondary to injury of the spinal cord from trauma and poliomyelitis. Histologic data are adapted from Minaire *et al.* (1974), and calcium balance data are from Heaney (1960).

unknown function (sialoprotein, osteocalcin, and osteonectin) by bone cells emphasizes the control that bone has over its own metabolism (Price and Baukol, 1980). A useful marker of bone collagen metabolism is hydroxyproline, formed by adding a hydroxyl group to the proline already incorporated into bone collagen (Prockop and Juva, 1965). When collagen breaks down, a small fraction of hydroxyproline (25%) is not hydrolyzed completely and is excreted in the urine. Because of this, urinary hydroxyproline can be used as an index of bone resorption in healthy adults on gelatin-free diets; however, one must be very cautious about using this index in growing children, since larger hydroxyproline-containing peptides associated with newly synthesized collagen in growing bone may also be present in the urine (Prockop and Kivirikko, 1967).

In each anatomic bone, collagen is mineralized in a highly ordered laminated fashion to form two general types of bone, trabecular and cortical. These bone types are distributed differently in the axial and appendicular skeleton and are partly responsible for the shape of the bones. Trabecular or cancellous bone, appropriately named for its spongy appearance, is located in the metaphyseal regions of the long bones and between cortical surfaces of flat bones. Compact or cortical bone forms the shaft of all long bones and the outer shell of flat bones. The amount of surface in relation to calcified matrix is greater in trabecular than in cortical bone. Thus, certain bones, e.g., vertebrae, which have a relatively larger amount of trabecular bone, are more vulnerable to the consequences of a change in the surface metabolic activities. In humans and other higher animals, a major component of cortical bone is associated with the haversian canal system consisting of concentric rings of calcified bone formed around endosteal capillaries. Bone is in a dynamic and constant state of change through growth or renewal, which may take place at its periosteal (outer membrane layer), endosteal (inner layer), or haversian canal surfaces. To distinguish these three bone surfaces which have different rates of growth and skeletal metabolism throughout life, Frost (1969) introduced the concept of bone cell envelopes. The periosteal bone cell envelope, which is active during general growth, slows down with aging so that the increase in the transverse diameter of long bones is barely measureable in humans by the sixth or seventh decade of life (Garn *et al.*, 1967). In contrast, the endosteal bone cell envelope and its extension, the haversian envelope, participate in the repair and renewal of tissue throughout life.

This process of bone renewal is termed "remodeling" and is to be distinguished from the process of modeling which is most active during growth. Modeling refers to the continuous process of growth (apposition) on one surface or of resorption on another surface; these activities result in alteration of the gross structure of bone (Parfitt, 1979). Remodeling refers to maintenance of bone tissue. An important feature of remodeling bone tissue is the different age and degree of mineralization of its discrete bone units. These units are most easily identified anatomically in haversian bone by a central canal (capillary) and a cement line around the calcified rings that separate osteons (haversian systems) from one another. Some units may last for 10 or more years, whereas others may last only a few months. These individual packets of bone are best thought of as metabolic units (Frost, 1963) to emphasize their intermittent active participation in maintaining the structural integrity of bone (skeletal homeostasis), as well as their continued function in the regulation of mineral metabolism (calcium homeostasis) (Rasmussen and Bordier, 1974).

The function of each bone metabolic unit (BMU) ultimately depends on the function and activity of the cells it contains. These are identifiable as osteocytes, osteoblasts (bone-forming cells), or osteoclasts (bone-resorbing cells). The osteocyte, the most common cell in bone, is essentially an osteoblast that has been buried within the layers of calcified matrix (Parfitt, 1984) and that retains communication with its former neighboring cells on the surface of a BMU by projections that give the cell a stellate appearance. This structural feature of bone is important to our understanding both of the survival of the osteocyte in hard tissue and of the continued participation of the osteocyte in the regulation of mineral metabolism long after the BMU has been structurally completed. The precursors of the osteoblast and osteocyte are thought to be mesenchymal cells lining the periosteal and endosteal resting surfaces (Owen, 1982), while the osteoclast is believed to originate from a precursor common to the monocytic phagocyte (Jee and Kimmel, 1977). Bone remodeling is initiated by these cells in each BMU in an ordered sequence of events under the influence of a variety of stimuli. The best known of these is parathyroid hormone (PTH), which activates the osteoclast to resorb calcified matrix. After a quiescent period, osteoblasts synthesize new matrix which is later calcified. Under normal conditions, these processes of resorption (which may take a month) and formation (which takes about 3 months) in some way are coupled to one another so that the structure of the skeleton and its integrity are maintained.

The physical or chemical signals that induce bone cells to increase activity in response to stimuli from exercise and loading remain unidentified. Recent work suggests that prostaglandin E_2 and/or the cyclic nucleotides are synthesized by the osteoblast in response to biomechanical stimuli (Rodan et al., 1975; Somjen, et al.,1980). Perhaps substances produced locally in response to physical forces enhance or modify remodeling activity and so change the general architecture of bone (Lanyon, 1985). How, or if, this process alters bone cell activity during bed rest or immobilization is unclear. Bassett and Becker (1962) demonstrated that bone could respond to pressure with a change in electrical current. A negative potential generated on the compressed side and a positive charge on the opposing bone surface could alter the local ionic composition and the activity of bone cells. The physiological importance of this electrical property of bone or other factors influencing bone cell activity is not known.

II. CURRENT CONCEPTS IN CALCIUM METABOLISM

Although the skeleton contains almost all the body's calcium, it is only one of three target organs that participate in the regulation of calcium homeostasis. The renal tubule, intestinal mucosa, and bone surfaces all contain polarized cells that are highly specialized with respect to calcium transport. These cells effectively modulate the transfer of calcium in and out of extracellular fluid to keep blood calcium remarkably constant. The direction of flow results from the interaction of three calcium regulating hormones (parathyroid hormone [PTH], calcitonin, and vitamin D) with the ions magnesium, phosphorus, sodium, hydrogen, and calcium itself. PTH and 1,25-dihydroxyvitamin D (well-known hypercalcemic hormones) favor the translocation of calcium to the extracellular fluids, whereas calcitonin (a hypocalcemic hormone) acts in the opposing direction (Rasmussen, 1971). These hormones also have other actions that affect calcium homeostasis. PTH stimulates osteoclastic resorption of bone, directly activates adenylate cyclase, decreases renal tubular reabsorption of both phosphate and bicarbonate and, indirectly, increases 1-alpha hydroxylation of 25-hydroxyvitamin D by the kidney. The vitamin D hormone (1,25-dihydroxyvitamin D) enhances the absorption of calcium and phosphorus from the intestine and the resorption of these minerals from bone. Other derivatives of vitamin D or the D hormone mineralize bone collagen and cartilage and have specific effects on calcium and phosphate transport in muscle and kidney. Calcitonin inhibits osteoclastic resorption of bone; decreases gastric acid secretion; increases the renal excretion of calcium, phosphate, and sodium; and increases the intestinal secretion of water, sodium, potassium, and chloride.

Calcium in blood is found primarily in plasma and is the smallest physiological compartment of calcium (0.08% of total body calcium). Its regulation is critically important to life. Even during growth, there is a steady relationship between blood and bone calcium exchange in the dog and chick (Klein, 1981). The two major calcium regulating hormones (PTH and vitamin D) respond to perturbations in blood calcium, so that a constant circulating level of calcium is maintained. A modest decrease in the calcium concentration of the blood is rapidly followed by an increase in the secretion and/or synthesis of PTH. The rise in PTH in turn acts directly on the kidney to conserve, and on the bone to mobilize, calcium and to restore the blood calcium level (held at 8.9–10.1 mg %) within a few hours (Lund et al., 1980). The effect of PTH on blood phosphorus, through phosphaturia, is just the opposite. Through the more delayed actions of 1,25-dihydroxyvitamin D, whose production is enhanced by a parathyroid-induced decrease in phosphorus and a stimulation of renal 1-alpha hydroxylase activity, intestinal absorption of calcium is increased. Conversely, an increase in serum calcium suppresses secretion of PTH. This reduces circulating calcium directly by inhibiting renal tubular reabsorption of calcium and decreasing osteoclastic resorption, and indirectly by favoring the renal retention of phosphate.

Calcitonin participates in the physiological regulation of calcium, but its role as an antagonist to PTH in the scheme just described is still unclear (Austin et al., 1979). Demonstration of responses of the calcium endocrine system to physiological changes are confounded by diurnal variation in all three calcium-regulating hormones (Hirsch and Hagaman, 1982; Jubiz et al., 1972; Markowitz et al., 1981), which does not appear to be directly related to the diurnal variation in blood calcium. In fact, even urinary excretion of calcium (normal range of 100–300 mg per 24 hr) has a circadian rhythm that peaks in the evening; this rhythm disappears in chronic disuse osteoporosis (Moore-Ede et al., 1972). These observations, coupled with the difficulty in measuring active, circulating peptides, may be one reason why expected changes in parathyroid hormone and calcitonin levels have not been observed in bed rest subjects (Arnaud et al., 1974). Four out of five subjects studied by Vernikos-Danellis and associates (1974) throughout

56 days of bed rest showed both an early increase in PTH which persisted and a greater 24-hr fluctuation than during the pre-bed-rest period.

An impressive feature of calcium metabolism is the way the calcium endocrine system modulates the exchange of large amounts of calcium between compartments (bone, soft tissues, and blood) that are quantitatively so different. All the calcium in blood (500 mg in an adult) is exchanged with an amount 10 times larger (the exchangeable calcium pool), without perceptible changes in the total blood calcium level. The continuous turnover of calcium, as determined by kinetic studies using radioactive calcium, involves a small fraction of the total body calcium, which is estimated to be approximately 100 mg/kg (Heaney and Whedon, 1958). In normal adults, slightly less than one-tenth of this exchangeable pool of calcium is thought to enter and leave bone (4–700 mg/day; Neer *et al.*, 1967). However, patients with metabolic bone disease dramatically illustrate the powerful influence of abnormal endocrine status and biomechanical forces on the movement of calcium in and out of bone. In surgical hypoparathyroidism, Heaney and Whedon (1958) and others found that calcium turnover decreased by 80%. Patients with disuse osteoporosis studied during the more acute hypercalciuric phase following paralysis from poliomyeltis showed increases in calcium turnover due primarily to net egress of calcium from bone (Heaney, 1960).

Since the integrity of the skeleton is maintained by the *coupling* of the processes of bone formation and resorption, a disturbance in one process *not* directly linked and coupled to the other will alter total skeletal mass. Kinetic data of calcium metabolism are difficult to reconcile with dynamic data of bone histology, even though we are aware that changes in the appearance of bone tissue reflect past events and that the calcium pools labeled in acute kinetic studies may reflect exchangeable calcium on bone surfaces. Nevertheless, both kinetic and dynamic techniques indicate that the demineralization of bone during bed rest occurs from an imbalance in which depressed formation is associated with a relative (Jowsey, 1971) or absolute (Lockwood *et al.*, 1975) increase in resorption. The very early increase in calcium accretion by bone, as observed from kinetic data, may actually be an acquisition of bone mineral by unmineralized osteoid synthesized prior to the inactivity (Heaney, 1960). Primary and secondary effects of the renal and intestinal handling of minerals, which may well account for the individual variation in the human response to bed rest, are not well documented.

III. NEGATIVE CALCIUM BALANCE DURING BED REST

The first long-term human bed rest study in otherwise normal individuals conducted by Whedon and co-workers in 1947 still remains one of the most informative records of the effects of immobilization on calcium metabolism (Dietrick et al., 1948). Metabolic changes were detailed while both diet and the degree of activity were strictly controlled. The four individuals studied consumed 850–920 mg of calcium and 1500–1640 mg of phosphorus daily. The proportion of carbohydrate, fat, and protein (63, 21, and 16%, respectively) was typical of American diets. Inactivity was assured by immobilizing the lower extremities and pelvic girdle in bivalved plaster casts that were removed only 30–40 min daily to allow excretory functions. Control data were acquired during 6 weeks of ambulation. Then the subjects were immobilized for 6–7 weeks. Although the responses of each individual varied, negative calcium balance occurred in all. Renal excretion of calcium, which doubled in each individual, accounted for the major portion of the calcium loss (ranging from 68 to 297 mg/day) while fecal calcium excretion showed both less loss and variation (ranging from 39 to 80 mg/day).

Donaldson and co-workers (19701) extended the duration of bed rest from 7 to 30 weeks in three subjects who were kept at bed rest unrestrained by plaster casts. These investigators demonstrated the pattern of change in calciuria over the entire 7-month period. Urinary calcium excretion accounted for most of the negative calcium balance during the first 3 months, ranging from 202 to 254 mg/day. Figure 2 shows how urinary calcium gradually increases to a maximum of 120 mg above ambulatory levels between the fifth and seventh week and then declines to about 50 mg above control at 18 weeks in the United States Public Health Service Hospital studies (Schneider et al., 1974). Gradual increases in the fecal excretion of calcium tended to account for 50 percent of the negative calcium balance observed *after* the eighth to tenth week of bed rest in contrast to the first few weeks. Normal pre-bed-rest calcium loss in the stool on an average diet was 750–800 mg per 24 hr and represented about 80% of the daily average intake.

IV. CHANGES IN BLOOD CHEMISTRIES DURING BED REST

Modest increases in the ionized fraction of calcium were documented in healthy young adult men immobilized by plaster casts during the first week of immobilization (Heath et al., 1972). As shown in Fig. 3, an increase in total serum calcium was recorded on the second day of immobilization, and serum phosphorus increased on the second and seventh day of horizontal bed rest (Leach et al., 1981). The absence of a corresponding increase in serum albumin suggests that this change is not related to hyperproteinemia secondary to fluid shifts and a contracted plasma volume. This change is probably transient, because weekly analyses performed by one of the authors (V.S.) during a 20-week bed rest study showed no evidence of a sustained increase in ionized calcium and no change in total serum calcium (see Fig. 4).

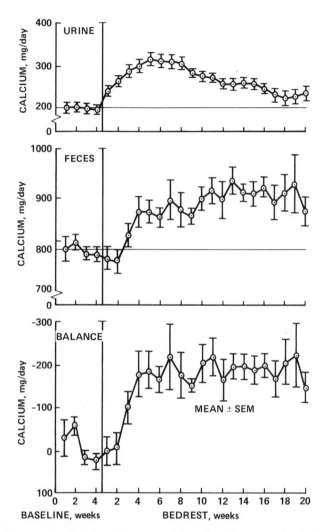

Fig. 2. The excretion of calcium in the urine and feces, in milligrams per day, in 12–32 healthy subjects immobilized by bed rest for 20 weeks. Dietary calcium was kept constant in these subjects at 1000 mg/day. Fecal calcium was corrected for time by polyethylene glycol 4000 administered three times daily in a dose of 500 mg. The lowest panel shows the results of the balance study estimated from the intake and excretion of calcium in the urine and feces.

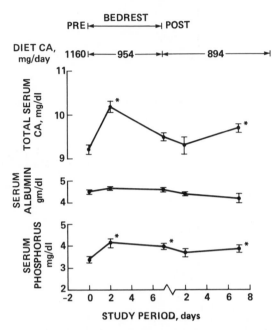

Fig. 3.   Total serum calcium, serum albumin, and serum phosphorus during the first week of bed rest in six healthy subjects. This study was conducted by Drs. H. Sandler, D. Goldwater, and C. Leach at the NASA-Ames Research Center in 1981 as part of the United States/U.S.S.R. Hypo-kinesia Study. The asterisks indicate the data points which are different ($p < .05$) from ambulatory control values (day 0).

The role of PTH per se in the hypercalcemia seen during immobilization still remains confusing. Normal, low, and elevated levels of PTH have been recorded in patients with hypercalcemia following paraplegia (Arnstein *et al.*, 1973; Chantraine *et al.*, 1979; Leman *et al.*, 1977; Stewart *et al.*, 1982). No consistent change in PTH was observed in healthy bed rested subjects. However, Stewart and associates (1982) found the expected decreases using midregion-specific radioimmunoassays for PTH and nephrogenous cyclic adenosine 3′,5′-mono-phosphate (AMP) as a measure of renal response to PTH in a study of 13 patients immobilized by traumatic spinal cord injury. During the hypercalciuric acute phase of immobilization (5–21 weeks after the injury), they found reduced excretion of urinary cyclic AMP and decreased serum PTH in association with serum calcium levels at the upper limit of normal; serum phosphorus was increased or high normal. In addition, plasma 1,25-dihydroxyvitamin D was reduced, even though dietary intake of calcium was normal. These findings do not support a conclusion that PTH acts on the skeleton to release calcium during immobilization. PTH may, however, through decreased reabsorption of calcium in the renal tubule, augment calciuria. The decrease in nephrogenous cyclic AMP would suggest reduced responsiveness of at least one target organ to PTH. The extent to which the parathyroids are involved in the local resorptive process in bone during immobilization remains unknown. Data that conflict with the findings of Stewart *et al.* (1982) in paraplegia and that suggest the paramount importance of PTH in resorption during disuse come from studies in dogs in which immobilization osteoporosis of casted limbs was prevented by parathyroidectomy. This information suggests that PTH, at least if not directly responsible for the loss of calcium from the skeleton, has a permissive effect (Burkhardt and Jowsey, 1967).

Theoretically, calcitonin, an inhibitor of bone resorption and an antagonist to PTH, would be an effective agent in reducing the resorptive process during immobilization. This has not proved to be the case in practice. Treatment with 100 Medical Research Council units of synthetic salmon calcitonin for 8 weeks was not successful in preventing the negative calcium balance and the loss of bone mineral in two subjects (Hantman *et al.*, 1973). In fact, calciuria increased during this treatment, and the calcium balance worsened. In two other subjects who were given calcium and phosphate supplements in addition to calcitonin and intermittent compression, no improvement was found in the negative calcium

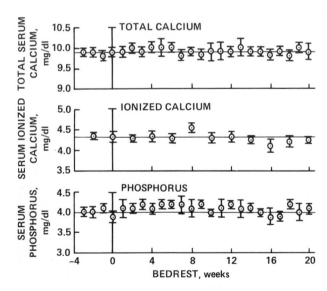

Fig. 4.   Total serum calcium, serum ionized calcium, and serum phosphorus in weekly measurements in 4–32 healthy subjects immobilized by bed rest at the U.S. Public Health Service Hospital in San Francisco, California.

balance. These results in humans and the failure of calcitonin to prevent bone
loss in experimental animals immobilized by nerve resection do not support the
view that calcitonin is a major factor in bone loss during disuse (Braddom *et al.*,
1973). Possible use of calcitonin, however, as a therapeutic agent for bone loss
has not been thoroughly investigated and may be dose dependent. Doses of
calcitonin that may be effective in preventing bone resorption may, through
hypocalcemia and increased excretion of calcium, aggravate negative calcium
balance and deliver conflicting signals to bone and negate the effectiveness of
calcitonin.

V. INTESTINAL CALCIUM ABSORPTION DURING BED REST

Available data suggest that intestinal calcium absorption is suppressed after 4
weeks of bed rest as shown in Fig. 5. Supplementation with additional dietary
calcium does appear to overcome the malabsorption of calcium and yet does not
excessively exaggerate the hypercalciuria of bed rest or immobilization (Mack
and LaChance, 1967). Negative calcium balance was improved by 1–2 grams of
dietary calcium by these investigators. Later, Hantman and associates (1973)
prevented negative calcium balance during the acute phase of bed rest with daily
supplements of dietary calcium (1.8–2.3 gm) and phosphate (3.0 gm of phosphorus). The supplements were given to four individuals for 8 weeks following
12 weeks of bed rest. Similar supplements given to a fifth subject during the first
8 weeks of bed rest did not totally reverse the already negative calcium balance in

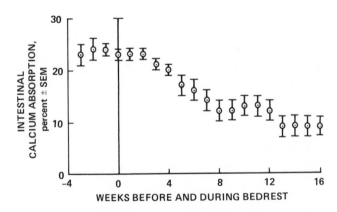

Fig. 5. The intestinal absorption of calcium estimated from balance data in 19–47 healthy
subjects immobilized by bed rest for 16 weeks. Absorption is determined in subjects consuming 1000
mg daily.

this subject. However, the effect of dietary calcium supplementation was not sustained. Calcaneal mineral density did decrease 10–30% in four subjects treated with mineral supplements between 12 and 20 weeks of bed rest, but urinary hydroxyproline excretion in four subjects during mineral supplementation was less than in the preceding bed rest period. Mack and LaChance (1967) determined the relationship between the level of dietary calcium and calcium balance during 14-day bed rest studies and found that the critical level of dietary calcium that reduced negative calcium balance to less than 7% of intake was 1000 mg per day. This amount increased urine calcium by 40 mg per day during ambulation but did not change the calcium excretion during bed rest. This intake is within the range of dietary calcium (1–2 gm) in which the efficiency of absorption is regulated by the vitamin D hormone, 1,25-dihydroxyvitamin D (Heaney et al., 1975). The correlation between circulating levels of this latter hormone and intestinal calcium absorption in healthy ambulatory adults is very high, no matter what the method for measuring absorption (Lemann et al., 1983). From estimates of balance data in bed rest subjects, calcium absorption is gradually reduced from 24 to 12% of intake during the first 8 weeks and then reaches a nadir which plateaus at 9% of ingested calcium after 12 weeks (Fig. 5). Results of intestinal calcium absorption in 6 paraplegics tested by Bergman and associates (1977) were variable. Stewart and co-workers (1982) reported consistently reduced levels of 1,25-dihydroxy-vitamin D in 13 paraplegic patients on a constant dietary calcium intake but did not measure intestinal absorption of calcium. Importantly, these investigators did not observe increases in urinary calcium in paraplegics following calcium supplements. In contrast to the effects of daily dietary calcium, Grigoriev (1981) found an increase in calciuria and reduced parathyroid endocrine levels following a single oral calcium lactate load in bed rest subjects (6 months), indicating a reduced capacity to retain a high dose of calcium probably absorbed through non-endocrine related systems. These data suggest that chronically in paraplegics and acutely in bed rest subjects, the endocrine regulation of dietary calcium is intact. Neither acute loads nor restricted intakes are warranted in these patients. All of the preceding evidence points to decreased efficiency in intestinal absorption of usual dietary intakes of calcium during immobilization. An adequate dietary intake of calcium, i.e., 0.8–1.5 gm daily, if tolerated by the patient, may be essential in preventing negative calcium balance.

VI. RENAL EXCRETION OF CALCIUM DURING BED REST

During bed rest, the absolute degree of calciuria approaches the upper limit of normal (350 mg per day in males). In 30 males studied by Schneider and associates (1974), the peak mean values during the fifth to seventh week of bed rest were 320 mg per day. Urine creatinine values remain relatively constant so that a slight increase in the ratio of urinary calcium to creatinine occurs. The slight increase in urinary pH observed by Deitrich and associates (1948) has not been a feature of the longer bed rest studies conducted at the Public Health Hospital in San Francisco (Schneider et al., 1974). Sodium and calcium are thought to share a common transport mechanism in the proximal tubule. During the first few days of bed rest, fluid shifts that contract the plasma volume and expand the extracellular fluid volume are associated with a diuresis (see Chapter 5). However, an increase in sodium excretion during the first 24 hr is associated with only a modest increase in calcium excretion (Greenleaf et al., 1977). Furthermore, sodium excretion does not increase beyond the fifth week, whereas calcium excretion does. Recent work as described in Chapters 2 and 5, explains that a change in sodium level in these instances may be governed by factors such as excretion of an atrial natriuretic factor and/or aldosterone from the adrenal cortex. Nevertheless, these different patterns of excretion do suggest that more than one renal mechanism for the tubular reabsorption of calcium is operating in bed rest subjects. As demonstrated by Griffith (1971), the calciuria of bed rest can be reduced by about 50% by either a low sodium diet that contracts the extracellular fluid volume, or by administering a thiazide diuretic (Costanzo and Weiner, 1974). Thiazide diuretics promote sodium excretion but have a hypocalciuric effect which is independent of PTH (Eknoyan, 1970). Rose (1966) attempted to reduce the calciuria and negative calcium balance in two patients by use of a thiazide compound. Urinary calcium excretion was reduced in both, but an increase in fecal calcium in one patient prevented any change in the calcium balance. Hypercalcemia, or at least an increase in serum calcium, is a common consequence of thiazide therapy and can be a major deterrant to its use during immobilization (Stote et al., 1972).

Negative phosphorus balance is also a characteristic feature of immobilization (Fig. 6). Since only 85% of body phosphorus is in bone, and phosphorus is a major intracellular anion of muscle, losses of this mineral have been attributed to muscle, as well as bone, atrophy. Phosphate supplements of 1–2 gm daily not only improve the negative phosphorus balance of immobilized patients but also improve the negative calcium balance, apparently by decreasing the calciuria associated with bed rest (Goldsmith et al., 1969). Phosphorus supplementation, given at the beginning of immobilization, did not appear to augment the increase in serum phosphorus noted in all subjects but did tend to decrease the level of serum calcium from the high-normal to midnormal ranges. Such findings suggest that phosphate supplements may increase bone formation and/or may inhibit bone resorption in the immobilized subject (Goldsmith et al., 1971). Deposition of calcium phosphate in soft tissues is an alternative explanation (Hebert et al., 1966).

The anabolic hormones, testosterone and estradiol, which can affect phosphate metabolism and bone growth, were given to patients immobilized by pol-

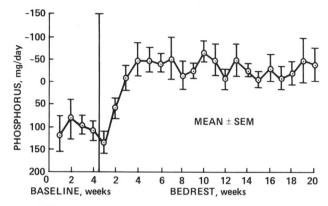

Fig. 6. Phosphorus balance in 12–32 healthy subjects immobilized by bed rest. Subjects consumed 1700 mg of phosphorus daily.

iomyelitis. Whedon and Shorr (1957) reported a decrease in mineral loss in these patients following administration of these hormones. Anabolic hormones may decrease urinary calcium rather than change bone metabolism per se. Calcium kinetic studies failed to demonstrate an increase in bone formation, but rather showed a reduced rate of bone resorption when the drugs were administered during the acute phase of disuse osteoporosis (Heaney, 1962).

VII. INFLUENCE OF AGE AND SEX ON CALCIURIA

Because children have relatively higher bone turnover than adults, one would expect them to show a more rapid and a greater loss of bone during disuse. Millard and associates (1970) studied 12 adolescents immobilized before and after spinal fusion for scoliosis and found that urinary calcium reached a maximum between 1 and 12 weeks, as it does in adults, and that the absolute amounts of calcium in urine were similar to those of adults. Only relative to body weight was the loss of bone mineral greater in the children. After 3–4 months the calciuria decreased in all children who were immobilized for 7–8 months. Hypercalcemia was not seen in any of these patients. However, Rosen and associates (1978) did detect transient increases in total serum calcium in 25% of his series of children with casts on one extremity. These same investigators also evaluated vitamin D status in another group of subjects and found no correlation between the nutritional state of vitamin D and any of the biochemical findings to incriminate vitamin D excess as a factor in hypercalcemia or bone loss. In a larger study involving 76 male and 64 female patients between the ages 14 and 57 years (Rose, 1966), the hypercalciuria of bed rest was less in men over 40 years old and in women over 75 years old than in younger patients. These apparent differences with age may reflect the decrease in bone turnover or calcium metabolism with age, or an effect of the underlying problems that caused these orthopedic patients to be hospitalized. They were a highly select population on free diets. Yet findings do indicate sex differences in urinary calcium excretion and a decreased efficiency of intestinal absorption of calcium with advancing age (Ireland and Fordtran, 1973).

VIII. THE EFFECT OF BED REST ON BONE DENSITY

Bone demineralization is regularly associated with immobilization, but data regarding the extent of osteoporosis in the different bones of the skeleton are insufficient to resolve questions that would separate the effects of weight bearing from more generalized bone loss secondary to disturbed mineral homeostasis (Meier, 1984). In the longest bed rest study that included calcium excretion, Vogel (1971) measured bone mineral content of the central os calcis from 12 to 36 weeks of bed rest. In three subjects he found decreases ranging from 25-45% of the 12-week value, which suggests that at least in this bone, the mineral continues to leave bone at a variable rate for up to 9 months. The maximum losses occur in the first 6 months (Krolner and Toft, 1983). These data and the absence of demineralization in the radius led Donaldson and his colleagues (1970) to conclude that the dissolution of bone is greater in weight-bearing bones than in the remainder of the skeleton. Radiographic techniques that are sufficiently sensitive and safe in terms of radiation exposure (Cohn *et al.*, 1978) to measure small changes in bone density have only recently been applied to other bones of bed rest subjects. Krolner and Toft (1983) measured the bone mineral content of the second to the fourth lumbar vertebra with dual photon (^{153}Gd) absorptiometry in 34 patients hospitalized for low backache due to protrusion of a lumbar intervertebral disc. Measurements at admission and after about 27 days of bed rest revealed an average decrease in bone mineral of 3.6% (0.9% per week); most mineral was restored after 4 months of ambulation. An important aspect of this study was the observation of mineral loss from the lumbar spine in patients with some activity. Patients walked to the bathroom and participated in an exercise program in the recumbent position during their period of therapeutic bed rest. These investigators emphasize that trabecular bone, because of its relatively greater surface area to total area compared with cortical bone, is likely to show changes more rapidly. Bone density measurements of the trabecular and cortical areas of the radius and ulna have not shown bone mineral loss in bed rested subjects who have decreased calcaneal mineral (Vogel, 1971). However, in 36 patients with spinal cord injuries resulting in either para- or quadriplegia, Norland-Cameron densitometry measurements of the distal radius (which assess trabecular bone) showed decreases, whereas measurements of the more proximal radius (which is primarily cortical bone) showed no differences (Griffith *et al.*, 1976). Osteopenia in paralyzed limbs is easily understood and is probably related to the degree of disuse and local factors. To document a more generalized disturbance in calcium metabolism that secondarily involves skeletal homeostasis requires further measurements in trabecular bone in nonparalyzed limbs of paraplegics.

IX. EFFECTS OF IMMOBILIZATION ON BONE COMPOSITION AND STRUCTURE

Hydroxyproline is a component of collagen that is excreted in the urine in quantities that reflect its turnover in bone. Hydroxyproline excretion parallels that of calcium and reaches a maximum increase of 10 mg per day after 4–6 weeks of bed rest, gradually declining to levels of 4 mg per day greater than baseline after the tenth week (Fig. 7). If one assumes that the urinary hydroxyproline represents the loss of bone matrix, then this measurement is an indirect estimate of bone resorption (Prockop amd Kivirikko, 1967). Such losses can be significant, as shown by the comprehensive studies of Klein and associates (1982). Dogs with one immobilized limb revealed a greater loss of collagen from

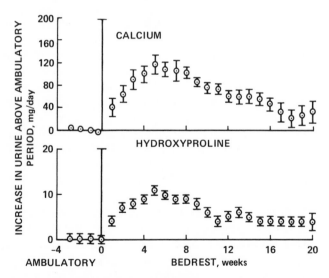

Fig. 7. The daily urinary excretion of hydroxyproline and calcium in the urine. Excretion is shown as the percentage of increase from the ambulatory control period of 4 weeks in 14–40 normal subjects during 20 weeks of bed rest.

bone (25–29%) than from soft tissue (8–10%) after 12 weeks (Klein *et al.*, 1982). The investigators analyzed bone resorption by prelabeling animals with [^{3}H]tetracycline (which is selectively deposited in bone forming at the mineraliz-·· ation front) and bone and soft tissue mass by chemical and isotopic content of calcium and collagen, respectively. The loss of collagen was greater in the tibia and femur than in the lateral meniscus or anterior cruciate ligament and was correlated with subsequent mechanical failure of bone. One-half the resorbed calcium appeared to be reutilized, whereas lost collagen was not replaced. These data support earlier kinetic studies of calcium metabolism, which predicted that formation of bone continues during disuse at a reduced rate, and that net loss of bone results primarily from increased bone resorption (Heaney, 1962).

Studies that localize the loss of bone to anatomic areas and their surfaces or envelopes have been carried out by Jaworski (1976) and Uhthoff and Jaworski (1978) in dogs. They demonstrated the importance of age in the activity of the skeletal response to immobilization. They immobilized one limb in a plaster cast for up to 40 weeks. Young animals showed the most rapid loss of bone. In young and old animals, the magnitude of bone loss was greatest in the distal bones of the casted limb; however, in older dogs the endosteal, rather than the periosteal, envelope was the main source of bone loss. In the more proximal bones of the extremity, the major area of loss was the periosteal envelope. The acute phase of bone loss lasted 12 weeks, and the transitional phase, 20 weeks, after which bone volume appeared to stabilize at about 50% of its initial mass (Uhthoff and Jaworski, 1978).

Jowsey (1971) studied bone turnover in bone biopsies obtained from 14 Mayo Clinic patients who were confined to bed rest for 4–17 days. All values for bone formation fell below the normal mean, whereas only 6 of 14 estimations of resorption surfaces were above the mean. Jowsey also analysed bone biopsies from 23 osteoporotic people at bed rest; 22 of 23 subjects showed increased areas of resorption by her measurement, yet formation values were clustered around the normal mean. She concluded that bed rest alone resulted in bone loss due to an imbalance in which reduced formation was associated with normal resorption while bed rested osteoporotics lost bone because of an increase in resorption.

More recent analysis of bone taken by iliac crest biopsy from 28 patients with paraplegia has provided cross-sectional and longitudinal quantitative data regarding changes in trabecular bone during the most severe form of acute disuse osteoporosis (Minaire *et al.*, 1974). Minaire and co-workers found an early increase in osteoclastic resorption surfaces and a later (after 12 weeks) increase in the size of the periosteocytic lacunae. Indices of bone formation (calcification rate and volume of osteoid tissue) were lower than normal for the first 12 weeks after the injury. After that time, osteoid volume was either within the low normal range or less than normal. The decrease in trabecular bone volume averaged 33% over 25 weeks and then appeared to stabilize just above the fracture threshold.

The increase in osteoclastic resorption surfaces that Minaire *et al.* observed is difficult to reconcile with the biochemical evidence of Stewart *et al.* (1982) of suppressed PTH activity, unless immobilization increases the sensitivity to circulating low levels of PTH or unless the resorption is a local phenomenon independent of PTH. It is also possible that suppression of PTH has adverse effects on bone formation directly or indirectly. Perhaps the suppression of bone formation lengthens the turnover rate for bone metabolic units; if so, resorption surfaces would fill in more slowly, and any histologic sample taken at such a stage would appear to overestimate bone resorption. In general, the duration of active bone loss in paraplegics appears to be transient. A new steady-state condition in which skeletal mass is reduced and the rate of bone turnover is low occurs in about 6–9 months. This interpretation agrees with observations of calcium kinetics in chronic disuse osteoporosis (Heaney, 1962), in which rates of exchangeable calcium are reduced and a positive calcium balance returns during the newly established steady state after a year or more.

X. EFFECTS OF EXERCISE ON BONE LOSS DURING BED REST

Since the most effective means of reducing the excretion of calcium in bed rest subjects proved to be quiet standing for at least 3 hr a day, Issekutz and associates (1966) attributed bone loss to an absence of weight bearing, rather than to a lack of activity per se. In a series of therapeutic trials from the Public Health Hospital in San Francisco, exercise with a pulley system and other exercise programs in the supine position actually aggravated negative calcium balance of young men and increased the excretion of urinary hydroxyproline during long-term bed rest (Schneider and McDonald, 1984). Neither isometric nor isotonic forms of exercise, 1 hr daily, altered the calciuria of bed rest (Greenleaf *et al.*, 1977). Intensive exercise, including trampoline exercises during bed rest, significantly reduced calciuria in Soviet studies, where a correlation between ion excretion and the level of exercise is reported (Grigoriev, 1981). These data partly seem to contradict the observations of greater bone mass in functioning than in nonfunctioning limbs of paralyzed patients (Stout, 1982), and the findings of less severe osteoporosis in paraplegics who walked with braces or crutches for 1 hr a day (Abramson and Delagi, 1961). These discrepancies could be explained by the remodeling sequence in which resorption precedes formation. Exercise may initially activate bone resorption and cause an early and temporary increase in urinary calcium coincident with the calciuria of inactive bed rest. More vigorous exercise may shift the direction of calcium flow to mineralized, newly formed bone more readily, or these phenomena may simply be explained by vascular changes. Although active exercise has proven beneficial in paraplegia, more passive forms of activity provided by head-up tilt on a table, rocking and oscillating beds, partial mobilization in a wheelchair, or walking for time periods shorter than an hour a day do not seem to be effective in preventing bone loss in the paralyzed patient, despite improved vasomotor tone (Wyse and Rittee, 1954).

In spite of the aforementioned observations in bed rest subjects, there is good evidence that bone mass can be increased by activity. This has been demonstrated by greater X-ray bone density in dominant arms of tennis players compared to their nondominant arm (Jones et al., 1977) and in the spine of women with postmenopausal osteoporosis who were placed on an exercise program for 2 years (Aloia et al., 1978; Krolner et al., 1983). The apparent discrepancy between balance studies and kinetic studies regarding the effects of bed rest on bone mineral density was pointed out by Lockwood and co-workers (1973). They noted marked individual variation in the degree of calcaneal mineral loss in bed rest subjects who showed a fairly consistent pattern in negative calcium balance. They were able to reconcile this problem and to predict bone loss when they related their data to the baseline excretion of urinary hydroxyproline. In the normal adult on a regular diet in which urinary hydroxyproline is a valid index of bone turnover, a finding of low turnover and high bone mass at the start of bed rest or immobilization is likely to be associated with less bone loss (Lockwood et al., 1973). Specific studies relating bone turnover and urinary hydroxyproline to exercise have yet to establish quantitative relationships among these variables. Studies that relate different types and amounts of exercise to bone mass will help resolve the many questions regarding the effects of exercise on bone loss during bed rest (Krolner et al., 1983; Marcus et al., 1985).

XI. BONE LOSS DURING SPACEFLIGHT

The similarity of metabolic changes in calcium metabolism in astronauts to those in bed rest subjects suggests that loss of bone that occurs during the acute phase of immobilization is dependent on gravitational loading (Smith et al., 1977). These changes occur independent of atmospheric pressure, which has been kept at near-earth levels in all recent Soviet and U.S. manned spaceflights. Lowered atmospheric pressure actually reduces calciuria (Lynch et al., 1967). Comprehensive physiological studies of mineral balance and bone density were conducted during the 28, 59, and 84 days of the Skylab missions. Negative calcium balance averaged 140 mg per day in the three crew members of the longest mission (Whedon et al., 1974, 1977). This negative calcium balance was associated with a shift in nitrogen balance to a loss of 3 gm daily and a negative phosphorus balance twice that of calcium. The maximum decrease in calcaneus density in two of the three 84–day crew was 7 and 4% of preflight values, with no corresponding changes in the radius when measured by photon absorptiometry (Smith et al., 1977). These crew members exhibited the largest negative shifts in calcium balance. Urinary hydroxyproline increased an average of 30% in some of the Skylab crew (Whedon et al., 1977) and was unchanged in three astronauts (Claus-Walker *et al.*, 1977). Individual variations in this parameter and in the documented decrease in calcaneal mineral density were considerable. The Soviets also have reported calcaneal mineral losses in eight cosmonauts who flew in missions lasting 75–185 days (Stupakov *et al.*, 1984). The decrease in calcaneal mineral density after landing has not exceeded 10% of preflight values in any of the individuals whose preflight measurement was used as baseline data. A 19.8% decrement in one cosmonaut after a 140-day flight can be discounted, because control measurements were taken 2 years after the flight. These relatively small losses of bone mineral were attributed to a strenuous in-flight exercise program, but systematic studies were not conducted with regard to calcium balance, kinetics, or control of individual exercise regimens. To date, the only direct measurements of weightlessness effects on bone metabolism have come from studies in rats, a species which does not normally have haversian systems in bone. Morphological data (bone histology) obtained in these animals over the course of three consecutive flights (Cosmos series) demonstrated reduced apposition of periosteal bone and no change in bone resorption (Morey-Holton and Wronski, 1981). In addition, calcium kinetics revealed some in-flight reduction in resorption, but these changes were far less than the reduction in formation (Cann and Adachi, 1983) and may have been related to an age-dependent decrease in resorption in young rats. In humans, increased in-flight excretion of urinary hydroxyproline and of calcium has provided indirect evidence of increased bone resorption (Rambaut and Johnston, 1979). However, measurements of PTH in astronauts showed no change (Leach and Rambaut, 1977). Future studies will be needed to understand both calcium and skeletal metabolism and their mechanism of change during spaceflight. Finally, in contrast to findings in long-term bed rest studies, the repair and return of bone mineral density in astronauts may not be assured. Tilton and co-workers (1980) found that bone mineral content of the calcaneus was still decreased in the Skylab crew 5 years after the flight. These findings must be interpreted cautiously due to the small number of subjects evaluated, possible age-related decreases in bone mineral content of the control group, and the lack of equivalent exercise histories for the two groups. This makes definitive interpretation of resultant data difficult. If the decrease in mineral content of bone during spaceflight is the result of underactive osteoblasts or overactive osteoclasts and postflight recovery is doubtful, there may be a fracture risk in later life (Parfitt, 1981).

XII. BONE CHANGES IN "BED RESTED" ANIMAL MODELS

The most commonly used animal models to simulate weightlessness or bed rest effects are the couched monkey and the suspended rat (Morey, 1979; Young *et al.*, 1983). Howard *et al.* (1971) immobilized monkeys (*Macaca nemestrina*) in a reclining position (couch) using cloth jackets for restraint. Unlike bed rested patients, the hydrostatic forces are not eliminated, but a chronic hypodynamic state results which unloads body musculature and long bones. The rat model places the long axis of the body in a 30°, head-down position, which unloads the rear limbs; the forelimbs are then used for body support, grooming, and moving the animals to food and water.

In monkeys, serial measurement of bone mass using dual-beam photon densitometry have detected changes only in the proximal tibia and not in the midtibia, ulna, or radius of monkeys after 6 months of couching. The major part of this loss occurs within the first 3 months of couching (Young and Schneider, 1981). In repeat studies significantly greater decreases in mineral density assessed by X-ray computer tomography were found in axial cancellous bone (lumbar vertebrae), as compared to appendicular cancellous bone (tibia) (Cann *et al.*, 1980). Spinal mineral density did not plateau as early as in the proximal tibia, suggesting that long-term immobilization may be more severe for the axial skeleton. Recently, similar measurements of the spine in humans have shown similar changes in patients bed rested for a dislocated intervertebral disc (Krolner and Toft, 1983). Direct histologic results of bone samples taken from the long-term monkey studies demonstrate resorption cavities with osteoclasts in the haversian envelope area adjacent to tendon attachments and muscle insertions. Increased osteoclastic activity can be noted in these areas and is associated with significant resorption of bone (Young *et al.*, 1983).

Findings from rat model studies conducted by Morey-Holton and co-workers (Globus *et al.*, 1984) have provided information on the effects of immobilization on skeletal metabolism in a species with a less complicated bone structure than the monkey. In the absence of haversian systems, the principal change in growing rats appears as decreased bone formation, without a corresponding change in bone resorption, only in the unweighted limbs (Morey, 1979). Measurable systemic effects in calcium homeostasis occur within a few days of immobilization and last only 5–7 days in animals with normal weight gain (Globus *et al.*, 1984). Depressed intestinal absorption of calcium does not occur in these partially immobilized animals. When nutrition is compromised, however, the period of reduced bone formation is extended. Thus, the extent and duration of bone changes during hind limb unweighting in the growing rat clearly appear to depend upon nutritional, physical and biochemical factors, as it does in humans. Unlike the effects in humans, the effects of endogenous steroids on bone metabolism may be a significant factor in the pathogeneses of disuse osteoporosis in restrained animals (Morey-Holton *et al.*, 1982).

XIII. SUMMARY

Figure 8 serves to summarize all the concepts discussed in this chapter related to bone and calcium homeostasis. The interaction of the intestine, kidney, and endocrine system with bone metabolism is stressed. Bed rest causes loss of both mineral and matrix from the skeleton. Both the spine and the more frequently studied calcaneus are involved. It is uncertain, however, whether these bones are affected because of their weight-bearing function or their component of trabecular bone. The finding that quiet standing for 3 or 4 hr per day prevents the loss

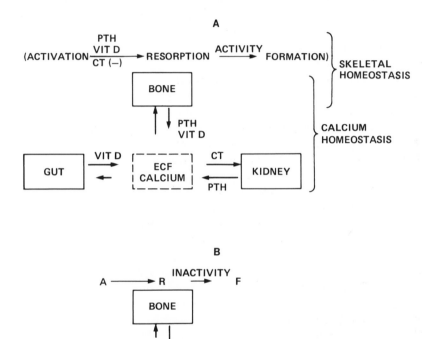

Fig. 8. Panel A depicts the flow of calcium in and out of the 3 target organs (gut, kidney and bone) through the extracellular fluid (ECF) in an active individual. The integrity of the skeleton is maintained by remodeling (activation of bone cells followed by resorptive and then formative processes), under the influence of biomechanical stimuli (activity) and hormones (parathyroid hormone (PTH), 1,25-dihydroxyvitamin D (VIT D), and calcitonin (CT). The hormones that have major effects on the regulation of the intestinal absorbtion of calcium (VIT D), the exchange of calcium in and out of bone (PTH and VIT D), and calcium excretion (CT and PTH) are shown next to the arrows in the direction of direct effects. Positive calcium balance is described. Panel B shows the changes in intestinal calcium absorbtion, bone accretion, and renal excretion of calcium which are thought to follow inactivity. The endocrine regulation or biomechanical signals which cause these changes is uncertain and not shown. Negative calcium balance is the effect of these theoretical shifts in calcium flow.

suggests that bed-rest-induced changes may be related more to physical factors associated with weight-bearing than to inactivity. Although muscular activity is known to stimulate bone formation, investigators as yet have not identified either the type of exercise or the duration needed to prevent bone loss in bed rested individuals. For ethical reasons bed rest studies in completely healthy subjects have been limited to durations of 5–6 months. The time required to resorb and to form a unit of bone is 3–4 months. Thus, the long-term adaptation of the skeleton to bed rest in healthy subjects is difficult to establish. From studies of paraplegics, immobilization appears to result in a negative calcium balance and losses of skeletal mass, which are self-limited. The acute phase appears to last about 3 months, during which time urinary calcium and hydroxyproline excretion are maximally increased, a pattern similar to that observed in long-term bed rest studies. After 3 months, bone dissolution continues but stabilizes at a lower rate. In paraplegia, one-third of trabecular bone volume is lost during the first 6–9 months following the causative spinal cord injury. Rates of bone loss for healthy bed rested subjects (estimated from calcaneal mineral density in 20-week studies) average 5% per month and are similar to findings in paraplegics. These data suggest that the magnitude of loss is unrelated to the underlying cause for immobilization. Age and bone turnover rate are critical factors in the response to inactivity. There is more rapid bone loss in younger individuals, with higher rates of bone turnover, than in older individuals, who tend to eventually lose a quantitatively similar amount of bone but do so more slowly.

The response of bone to immobilization not only appears to be an adaptive phenomenon of this organ system to the absence of weight bearing and to a change to the horizontal position, but also seems to be transient. Until all components of the response are understood, it is difficult to predict the duration of the acute and transitional phases that lead to a new steady state or the degree of involved bone loss. The mechanisms of bone loss during bed rest will not be defined until we understand what signals are generated in bone in response to changes in position and/or weight bearing, or both, and how these signals interrelate with the calcium endocrine system and other components of calcium metabolism as shown in Fig. 8.

REFERENCES

Abramson, A.S., and Delagi, E.F. (1961). *Arch. Phys. Med. Rehab.* **42**, 147.
Albright, F., Burnett, C.H., Cope, O., and Parsons, W. (1941). *J. Clin. Endocrin.* **1**, 711-714.
Aloia, J.F., Cohn, S.H., Ostuni, J.A., Cane, R., and Ellis, K. (1978). *Ann. Int. Med.* **89**, 356-358.
Arnaud, C.D., Goldsmith, R.S., Bordier, P.J., and Sizemore, G.W. (1974). *Am. J. Med.* **56**, 785-793.
Arnstein, A.R., McCann, D.S., Blumenthal, F.S., and Pruhty, J. (1973). *In* "Exerpta Medica,

Amsterdam, 1973" (B. Frame, A.M. Parfitt, and H. Duncan, eds.). Intl. Congr. Series No. 270, pp. 253-256.

Austin, L.A., Heath III, H., and Go, V.L.W. (1979). *J. Clin. Invest.* **64**, 1721-1724.

Bassett, C.A.L. and Becker, R.O. (1962). *Science* **137**, 1063-1064.

Bergmann, P., Heilporn, A., Schoutens, A., Paternot, J., and Tricot, A. (1977). *Paraplegia* **15**, 147-159.

Braddom, R.L., Erickson, R., and Johnson, E.W. (1973). *Arch. Phys. Med. Rehab.* **54**, 170.

Burkhardt, J.M., and Jowsey, J. (1967). *Endocrinology* **81**, 1053-1062.

Cann, C.E., and Adachi, R.R. (1983). *Am. J. Physiol.* **13**, R327-R331.

Cann, C.E., Genant, H.K., and Young, D.R. (1980). *Radiology* **134**, 525-529.

Chantraine, A., Heynen, G., and Franchimont, P. (1979). *Calcif. Tiss. Intl.* **27**, 199-204.

Claus-Walker, J., Singh, J., Leach, C.S., Hatton, D.V., Hubert, C.W., DiFerrante, N. (1977). *J. Bone Jt. Surg.* **59-A**, 209-212.

Cohn, S.H., Aloia, J.F., and Letteri, J. (1978). *Calcif. Tiss. Res.* **26**, 1-3.

Costanzo, L.S., and Weiner, I.M. (1974). *J. Clin. Invest.* **54**, 628-637.

Cowin, S.C., Lanyon, L.E., and Rodan, G. (1984). *Calcif. Tiss. Intl.* **36**, S1-6.

Deitrick, J.E., Whedon, G.D., and Shorr, E. (1948). *Am. J. Med.* **4**, 3-36.

Donaldson, C.L., Hulley, S.B., Vogel, J.M., Hattner, R.S., Boyers, J.H., and McMillan, D.E. (1970). *Metabolism* **19**, 1071-1084.

Eknoyan, G., Suki, W.N., Martinez-Maldonado, M. (1970). *J. Lab. Clin. Med.* **76**, 257-268.

Frost, H.M. (1963). "Bone Remodeling Dynamics." Springfield, Thomas.

Frost, H.M. (1969). *Calcif. Tiss. Res.* **3**, 211.

Garn, S.W., Rohmann, C.G., and Wagner, B. (1967). *Fed. Proc.* **26**, 1729-1738.

Globus, R.K., Bikle, D.D., and Morey-Holton, E. (1984). *Endocrinology* **114**, 2264-2270.

Goldsmith, R.S., Killian, P., Ingbar, S.H., and Bass, D.E. (1969). *Metabolism* **18**, 349-368.

Goldsmith, R.S., Richards, R., Dube, W.J., Hulley, S.B., Holdsworth, D., and Ingbar, S.H. (1971). *In* "Phosphate et Metabolisme Phosphocalcique." Laboratories Sandoz, Paris, pp. 271-291.

Greenleaf, J.E., Bernauer, E.M., Juhos, L.T., Young, H.L., Morse, J.T., and Staley, R.W. (1977). *J. Appl. Physiol.* **43(1)**, 126-132.

Griffith, D.P. (1971). *Aerosp. Med.* **42**, 1322-1324.

Griffiths, H.J., Bushueff, B., and Zimmerman, R.E. (1976). *Paraplegia* **14**, 207-212.

Grigoriev, A.I. (1981). *Acta Astronautica* **8**, 987-993.

Hantman, D.A., Vogel, J.M., Donaldson, C.L., Friedman, R., Goldsmith, R.S., and Hulley, S.B. (1973). *J. Clin. Endocrinol. Metab.* **36**, 845-858.

Heaney, R.P. (1960). *J. Lab. Clin. Med.* **56**, 825.

Heaney, R.P. (1962). *Am. J. Med.* **33**, 188-200.

Heaney, R.P., and Whedon, C.D. (1958). *J. Clin. Endocrin.* **18**, 1246.

Heaney, R.P., Seville, P.D., and Recker, R.R. (1975). *J. Lab. Clin. Med.* **85**, 881-890.

Heath, H. III, Earll, J.M., Schaaf, M., Piechocki, J.T., and Li, T-K. (1972). *Metabolism* **21**, 633-640.

Hebert, L.A., Lehmann, J., Peterson, J.R., and Lennon, E.J. (1966). *J. Clin. Invest.* **45**, 1886.

Hirsch, P.F., and Hagaman, J.R. (1982). *Endocrinology* **110**, 961-966.

Howard, W.H., Parcher, J.W., and Young, D.R. (1971). *Lab. Animal Sci.* **21**, 112-117.

Ireland, P. and Fordtran, J.S. (1973). *J. Clin. Invest.* **52**, 2672.

Issekutz, B., Jr., Blizzard, J.J., Birkhead, N.C., and Rodahl, K. (1966). *J. Appl. Physiol.* **21**, 1013-1020.

Jaworski, Z.F.G. (1976). *In* "Proceedings of the First Workshop on Bone Histomorphometry" (Z.F.G. Jaworski, ed.), University of Ottawa Press, pp. 254-256.

Jee, W.S.S. and Kimmel, D.B. (1977). *In* "Bone Histomorphometry: 2nd International Workshop, Armour Montagu, Paris," (P.J. Meunier, ed.).

Jones, H.H., Priest, J.D., Hayes, W.C., Tichenor, C.C., and Nagel, A. (1977). *J. Bone Joint Surg.* **59A**, 204-208.

Jowsey, J. (1971). *In* "Hypogravic and Hypodynamic Environments" (R.H. Murray and M. McCally, eds.), NASA SP-269, p. 111-119.

Jubiz, W., Canterbury, J.M., Reiss, E., and Tyler, F.H. (1972). *J. Clin. Invest.* **51**, 2040-2046.

Klein, L. (1981). *Science* **214**, 190-193.

Klein, L., Player, J.S., Heiple, K.G., Bahniuk, E., and Goldberg, V.M. (1982). *J. Bone Joint Surg.* **64A**, 225-230.

Krolner, B., and Toft, B. (1983). *Clin. Sci.* **64**, 537-540.

Krolner, B., Toft, B., Nielsen, S.P., and Tondevold, E. (1983). *Clin. Sci.* **64**, 541-546.

Lanyon, L.E. (1984). *Calcif. Tiss. Int.* **36**, S56-61.

Leach, C.S., and Rambaut, P.C. (1977). *In* "Biomedical Results from Skylab" (R.S. Johnston and L.F. Dietlein, eds.), NASA SP-377, pp. 204-216.

Leach, C.S., Goldwater, D., Sandler, H. (1981). Unpublished data.

Leman, S., Canterbury, J.M., and Reiss, E. (1977). *J. Clin. Endocrinol. Metab.* **45**, 425-428.

Lemann, J., Adams, N.D., Gray, R.W. (1983). *Mineral Electrolyte Metab.* **9**, 55.

Lockwood, D.R., Lammet, J.E., Vogel, J.M., and Hulley, S.B. (1973). *In* "Clinical Aspects of Metabolic Bone Disease" (B. Frame, A.M. Parfitt, and H. Duncan, eds.). Exerpta Medica, Amsterdam.

Lockwood,D.R.,Vogel, J.M., Schneider, V.S., and Hulley, S.B. (1975). *J. Clin. Endocrinol. Metab.* **41**, 533-541.

Lund, B., Sorensen, O., Lund, B., Bishop, J.E., and Newman, A.W. (1980). *J. Clin. Endocrin. Metab.* **50**, 480-484.

Lynch, T.N., Jensen, R.L., Stevens, P.M., Johnson, R.L., and Lamb, L.E. (1967). *Aerosp. Med.* **38**, 10.

Mack, P.B., and LeChance, P.L. (1967). *Am. J. Clin. Nutr.* **20**, 1194-1205.

Marcus, R., Cann, C., Madvig, P., Minkoff, J., Goddard, M., Bayer, M., Martin, M., Gaudiani, L. Haskell, W., and Genant, H. (1985). *Ann. Int. Med.* **102**, 158-163.

Markowitz, M., Rotkin, L., and Rosen, J.F. (1981). *Science* **213**, 672-674.

Meier, D.E., Orwoll, E.S, and Jones, J.M. (1984). *Ann. Int. Med.* **101**, 605-612.

Millard, F.J.C., Nassim, J.R., and Woolen, J.W. (1970). *Arch. Dis. Child.* **45**, 399.

Minaire, P., Meunier, P., Edouard, C., Bernard, J., Courpron, P., and Bourret, J. (1974). *Calcif. Tiss. Res.* **17**, 57-73.

Moore-Ede, M.C., Faulkner, M.H., and Tredre, B.E. (1972). *Clin. Sci.* **42**, 433-445.

Morey, E.R. (1979). *Bioscience* **29**, 168-172.

Morey, E.R., and Baylink, D.J. (1978). *Science* **201**, 1138.

Morey-Holton, E.R., and Wronski, T.J. (1981). *Physiologist* **24**, S45-48.

Morey-Holton, E.R., Bomalaski, M.S., Enayati-Gordon, E., Gonsalves, M.R., and Wronski, T.J. (1982). *Physiologist* **25**, S145.

Neer, R., Berman, M., Fisher, L., and Rosenberg, L.E. (1967). *J. Clin. Invest.* **46**, 1364-1379.

Neuman, M.W. (1982). *Calcif. Tiss. Int.* **34**, 117-120.

Neuman, W.F. and Ramp, W.K. (1971). *In* "Cellular Mechanisms for Calcium Transfer and Homeostasis" (E.G. Nichols, Jr. and R.H. Wasserman, eds.), pp. 197-209, Academic Press, New York and London.

Owen, M. (1982). *In* "Factors and Mechanisms Influencing Bone Growth" (A.D. Dixon and B.G. Sarnat, eds.), p. 19, Alan R. Liss, New York.

Parfitt, A.M. (1979). *Calcif. Tiss. Int.* **28**, 1-5.

Parfitt, A.M. (1981). *Acta Astronautica* **8(9-10)**, 1083-1090.

Parfitt, A.M. (1984). *Calcif. Tiss. Int.* **36**, S37-45.

Price, P.A. and Baukol, S.A. (1980). *J. Biol. Chem.* **255**, 11660-11663.

Prockop, D.J. and Juva, K. (1965). *Proc. Natl. Acad. Sci. USA* **53**, 661-668.

Prockop, D.J. and Kivirikko, K.I. (1967). *Ann. Intern. Med.* **66**, 1243-1266.

Quelch, K.J., Melick, R.A., Bingham, P.J., and Merueri, S.M. (1983). *Arch. Oral Biol.* **28**, 665-674.

Rambaut, P.C. and Johnston, R.S. (1979). *Acta Astronautica* **6**, 1113-1122.

Rasmussen, H. (1971). *Am. J. Med.* **50**, 567-588.

Rasmussen, H., and Bordier, P. (1974). "The Physiological and Cellular Basis of Metabolic Bone Disease." pp. 1-7, Williams and Wilkins Co., Baltimore, Maryland.

Rodan, G.A., Bourret, L.A., Harvey, A., and Mensi, T. (1975). *Science* **189**, 467-469.

Rose, G.A. (1966). *Brit. J. Surg.* **53**, 769.

Rosen, J.F., Udin, D.A., and Finberg, L. (1978). *Am. J. Dis. Child.* **132**, 560-564.

Schneider, A.B. and Sherwood, L.M. (1974). *Metabolism* **23**, 975-1007.

Schneider, V.S., and McDonald, J. (1984). *Calcif. Tiss. Int.* **36**, S151-154.

Schneider, V.S., Hulley, S.B., Donaldson, C.L., Vogel, J.M., Rosen, S.N., Hantman, D.A., Lockwood, D.R., Reid, D., Hyatt, K.H., and Jacobson, C.B. (1974). "Prevention of Bone Mineral Changes Induced by Bedrest: Modification by static compression, simulated weight bearing, combined supplementation of oral calcium and phosphate, calcitonin injections, oscillating compression, the oral diphosphonate disodium etidronate, and lower body negative pressure" (Final Report.) NASA CR-141453, Public Health Service Hospital, San Francisco, CA. NTIS No. N75-14431/1ST.

Seeman, E., Wahner, H.W., Offord, K.P., Kumar, R., Johnson, W.J., and Riggs, B.L. (1982). *J. Clin. Invest.* **69**, 1302-1309.

Smith, M.C., Jr., Rambaut, P.C., Vogel, J.M., and Whittle, M.W. (1977). *In* "Biomedical Results from Skylab" (R.S. Johnston and L.F. Dietlein, eds.), NASA SP-377, pp. 183-190.

Somjen, D., Binderman, I., Berger, E., and Harrell, A. (1980). *Biochim. Biophys. Acta* **629**, 91-100.

Stewart, A.F., Adler, M., Byers, C.M., Segre, G.V., and Broadus, A.E. (1982). *New Engl. J. Med.* **306**, 1136-1140.

Stote, R.M., Smith, L.H., Wilson, D.M., *et al.* (1972). *Ann. Int. Med.* **77**, 587-591.

Stout, S.D. (1982). *Calcif. Tiss. Int.* **34**, 337-342.

Stupakov, G.P., Kazeykin, V.S., Koslovskiy, A.P., and Korolev, V.V. (1984). *Space Biol. Med.* **18(2)**, 42-47.

Teitlebaum, S.L., Bergfield, M.A., Freitag, J., Hruska, K.A., and Slatopolsky, E. (1980). *J. Clin. Endocrinol. Metab.* **51**, 247-251.

Tilton, F.E., Degioanni, J.J., and Schneider, V.S. (1980). *Aviat. Space Environ. Med.* **51**, 1209-1213.

Uhthoff, H.K. and Jaworski, Z.F.G. (1978). *J. Bone Joint Surg.* **60B**, 420-429.

Vernikos-Danellis, J., Winget, C.M., Leach, C.S., and Rambaut, P.C. (1974). "Circadian Endocrine and Metabolic Effects of Prolonged Bedrest: Two 56-Day Bedrest Studies." NASA-TMX-3051.

Vogel, J.M. (1971). *In* "Hypogravic and Hypodynamic Environments" (R.H. Murray and M. McCally, eds.). NASA SP-269, pp. 261-269.

Walser, M. (1961). *Am. J. Physiol.* **200**, 1099.

Whedon, G.D. and Shorr, E. (1957). *J. Clin. Invest.* **36**, 941.

Whedon, G.D., Lutwak, L., Reid, J., Rambaut, P., Whittle, M., Smith, M., and Leach, C. (1974). *Trans. Assoc. Am. Phys.* **137**, 95-110.

Whedon, G.D., Lutwak, L., Rambaut, P.C., Whittle, M.W., Smith, M.C., Jr., Reid, J., Leach,

C.S., Stadler, C.R., and Sanford, D.D. (1977). *In* "Biomedical Results from Skylab" (R. Johnston and L. Dietlein, eds.). NASA SP-377, pp. 164-174.

Wyse, D.M. and Rittee, C.J. (1954). *Am. J. Med.* **10**, 645-661.

Young, D.R., and Schneider, V.S. (1981). *Calcif. Tiss. Int.* **33**, 631-639.

Young, D.R., Niklowitz, W.J., and Steele, C.R. (1983). *Calcif. Tiss. Int.* **35**, 304-308.

4

Effects of Inactivity on Muscle

HAROLD SANDLER
Cardiovascular Research Office
Biomedical Research Division
National Aeronautics and Space Administration
Ames Research Center
Moffett Field, California 94035

I. MUSCLES AND GRAVITY

Our ability to stand upright has not been accomplished without cost and has been made possible only through the development of an antigravity muscle system. These muscles are concentrated in the lower back, neck, abdomen, buttocks, and thighs and allow an erect posture, support the head over the spinal column, and center the body over the pelvis and legs. Upon lying down, this system is no longer needed, and the body shifts its center of gravity from the midpelvis to the middle of the back and lower chest as the body becomes supported over its entire length rather than by the legs and the feet. Furthermore, as pointed out by Browse (1965), there have been other body developmental changes that make lying supine both possible and comfortable, since we are one of the few creatures who can sleep on our backs. This results from the unique chest configuration and the small spinous processes of the vertebral bodies. When supine muscles other than the antigravity muscles take over important support roles for the head, shoulder girdle, and back, they assume functions that are not their usual custom. That this occurs becomes clinically obvious from symptoms of all bed rested subjects, confined for periods of 24–72 hr, who complain of generalized muscle aches and tired feelings, particularly in the low back and less frequently in the neck, arms, and legs, which can only be relieved by a change in body position.

II. FUNCTIONAL ANATOMY

Skeletal muscles constitute 40% of body mass and are responsible not only for body posture, but also for all body locomotion and useful physical work. Over

77

600 muscles can be identified in the body. Some are very small and consist of only a few hundred fibers; larger muscles may contain several hundred thousand fibers. These fibers are arranged in bundles, or fasciculi, whose functioning units are myofibrils and myofilaments, as shown in Fig. 1. All are embedded in connective tissue (primarily collagen) that runs from one end to the other of the muscle and divides fibers into muscle bundles. Viewed under the microscope, each fiber shows striation due to the presence of actin and myosin, which are the proteins responsible for subsequent muscle shortening. This process results when actin and myosin, after appropriate activation, are combined to form actomyosin (see Fig. 2). This property allows muscle to shrink to one-half its normal length. Muscle relaxation, which is also an active process, results when these muscle proteins are returned to their original separate actin and myosin resting state. The

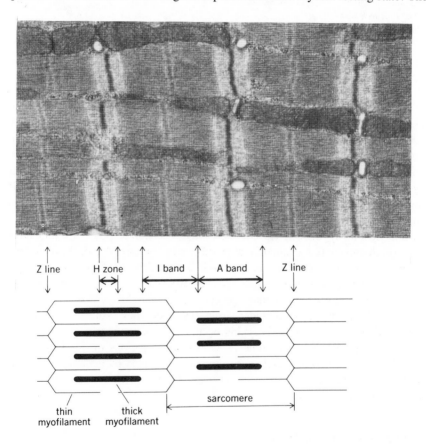

Fig. 1. Electron micrograph showing three myofibrils in a single muscle fiber. Diagrammed below is the organization of thick and thin myofilaments in a myofibril, which gives rise to the banding pattern in striated muscles.

work done by a muscle in a complete contraction increases with its length, one of the classic observations in muscle physiology. Resting length is about 60% of the level where maximum tension is developed. When stretched to about 175% of this length or beyond, force is no longer developed.

The total force of any given muscle represents the sum of that developed by its contained fibers. A muscle 1 cm^2 in cross section can maximally support a weight of 3–4 gm. Thus, an individual's overall strength inherently depends, in the first instance, on the cross-sectional area of his or her muscles. Almost all muscles are related anatomically and physiologically to the skeleton to which they are attached, although a few are attached from bone to skin or from one part of the skin to another, the latter occurring primarily in the face for speech and the expression of emotion.Movement of each body part is accomplished by close interaction of muscles, usually situated in opposing groups to cause physical displacement of a limb, the head, or the torso. Such actions depend totally on central nervous system coordination and processing of required information. Neither bed rest nor weightlessness alters these mechanical functions. However, the force that is needed or developed to perform any task, the interaction between opposing muscle groups, and the body support functions are markedly altered so as to reduce such functions during weightlessness or with inactivity. Therefore, the associated metabolic demand for accomplishing any given task at rest or when active is decreased and in most cases markedly so.

III. BIOCHEMISTRY OF MUSCLE CONTRACTION

All types of muscle—skeletal, cardiac, and smooth muscle (the latter primarily located in the walls of blood vessels and the gut)—are chemomechanical transducers converting chemical energy to mechanical energy and heat. This process takes place within the mitochondria located in every myofibril (see Figs. 1 and 2). The required energy source to initiate and sustain the reaction is supplied by adenosine triphosphate (ATP), which breaks down into adenosine diphosphate (ADP) and phosphate (PO_4^{2+}). This breakdown, which releases 7.6 kcal of energy per mole of ATP, allows actin and myosin to combine to form the active actomyosin complex resulting in muscle shortening, as follows:

$$ATP + actin + myosin \xrightarrow{Ca^{2+}} actomyosin + PO_4 + ADP + heat \qquad (1)$$

It is important to note that ionic calcium (Ca^{2+}) is required to enable the process. The exact steps involved in this latter process are shown in Fig. 2A (a through d). Tension is developed by change in cross bridges (Fig. 2B, b) due to formation of meromyosin (Fig. 2B, a) from light and heavy chain myosin (Fig. 2B, c), which in turn forms thick and thin myofilaments (see also Fig. 1). Energy for sustained or repeated contractions results from the breakdown of glucose and glycogen carbohydrates stored in the muscle. Fats and amino acids may also be used, but

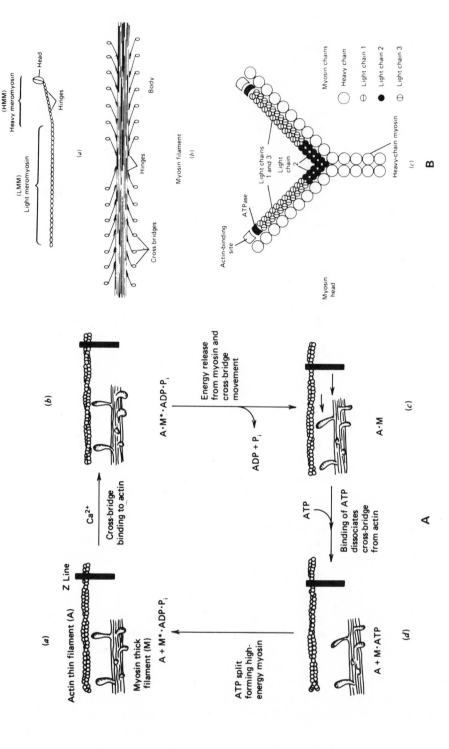

(HMM)
Heavy meromyosin

Head

Hinges

(LMM)
Light meromyosin

(a)

Hinges

Cross bridges

Body

Myosin filament

(b)

Myosin chains

○ Heavy chain
⊖ Light chain 1
● Light chain 2
⊕ Light chain 3

Light chains 1 and 3

Light chain 2

ATPase

Actin-binding site

Myosin head

Heavy-chain myosin

(c)

B

Z Line

Actin thin filament (A)

Myosin thick filament (M)

(a)

A + M* · ADP · P_i

Ca²⁺
Cross-bridge binding to actin

(b)

A · M* · ADP · P_i

Energy release from myosin and cross-bridge movement

ADP + P_i

A · M

(c)

Binding of ATP dissociates cross-bridge from actin

ATP

ATP split forming high-energy myosin

A + M · ATP

(d)

A + M* · ADP · P_i

A

their efficiency is far less. Since neither oxygen nor carbohydrates are always readily available, nonoxidative sources for energy release are also available and depend on phosphocreatine (CP) and nonesterified fatty acids (NEFA). These processes are used particularly during the initial phases of muscle contraction or during periods of high sustained exercise. High levels of effort cannot be sustained for long periods without sufficient oxygen supply, such a lack is indicated by the release of lactic acid, the main by-product of anaerobic metabolism, into the blood. In a maximal physical effort on a treadmill about 100 gm of lactic acid may be produced in a few minutes; in a well trained-athlete it may be higher, and blood concentrations rise from resting levels of 1 mmol/liter to 10-20 mmol/liter. It may take a considerable period for ATP, CP, and depleted muscle glycogen to be restored following an intense bout of exercise, even though muscle contraction is no longer taking place. This interval is termed "oxygen debt" and is related directly to the efficiency of the overall biochemical process, which is only 20%, the remaining 80% being lost as heat.

IV. NEURAL CONTROL

Skeletal muscle functions by means of a motor unit that consists of an anterior horn cell in the spinal column and its efferent fibers (alpha nerve fibers). Such fibers terminate in their respective muscles at the myoneural junction enabling transmission of nerve impulses from the spinal column and central nervous system to the muscles. The exact innervation of the muscle depends on its function and whether the fiber contracts slowly (Type I fiber) or fast (Type II fiber). Type II fibers reach peak force and contract two to four times faster than Type I fibers. Both fiber types occur in all muscles, with predominance of either Type I or Type II fibers depending on the type of activity required of the specific muscle. Figure 3 lists the distribution of these fibers determined by muscle biopsy from normal individuals and from those engaging in various kinds of athletic endeavors. Type I fibers predominate in weight lifters, whereas Type II

Fig. 2. Structure of myosin and its role in muscle contraction and relaxation. [from Brooks and Fahey, 1984, with permission.] **(A)** The cyclic process of muscle contraction and relaxation. In the resting state (a), actin (A) and energized myosin ($M^* \cdot ADP \cdot P_i$) cannot interact, because of the effect of tropomyosin. Upon release of Ca^{2+} from the sarcoplasmic reticulum, actin binds to energized myosin (b). Tension is developed and movement occurs with the release of ADP and P_i (c). Dissociation of actin and myosin requires the presence of ATP to bind myosin and to pump Ca^{2+} into the SR (d). Myosin is energized upon return to the resting state (a). **(B)** Myosin molecules have structural (light) and enzymatic (heavy ends) (a). The light meromyosin provides a connecting link to similar units in the myosin filament, which appear as a double-ended bottle brush (b). The head of heavy meromyosin contains binding sites for actin and ATP. Myosin in fast-contracting muscle has a higher ATPase activity than does myosin in slow muscle. These differences in ATPase activity are due to differences in light- and heavy-chain composition in the myosin head (c).

82   **Harold Sandler**

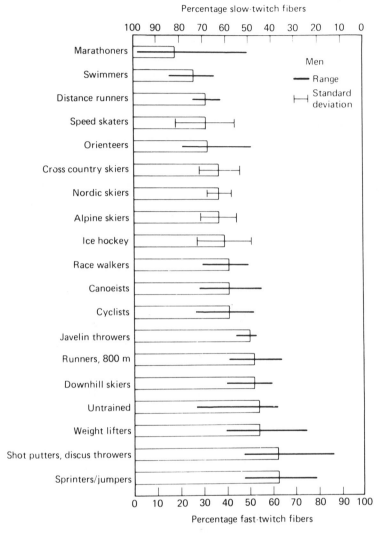

Fig. 3. Distribution of fast- and slow-twitch fibers in selected athletes. [From Brooks and Fahey, 1984.]

are more prominent in distance runners. Slow twitch Type I fibers predominate in the antigravity muscles. Electrostimulation performed over long periods can change Type II fibers into Type I (Riley and Allin, 1973). Transnervation experiments have also revealed that Type I fibers can be changed into Type II. Figure 3 clearly demonstrates the effects of physical work and exercise on the distribution of fiber types.

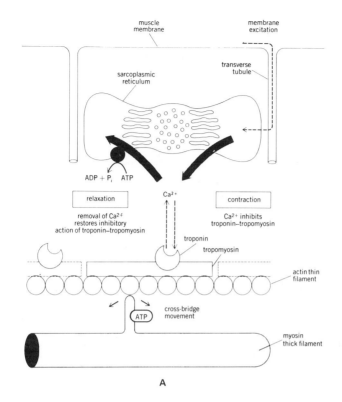

A

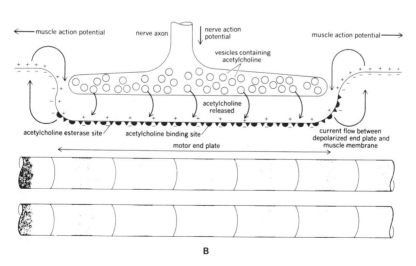

B

Fig. 4. Factors governing muscle excitation-contraction control. (A) Role of calcium in muscle excitation-contraction coupling. (B) Events occurring at a neuromuscular junction which led to action potential in the muscle membrane. [From Vander *et al.*, 1970.]

Figure 4 illustrates the stages or steps in the transmission of nerve signals to muscle. This process is termed excitation–contraction–coupling. It proceeds as follows: the nerve impulse coming down the nerve and arriving at the end plate releases acetylcholine (ACH) at the neuromuscular junction (see Fig. 4B). Acetylcholine is the chemical mediator that then conducts the impulse across the synapse to stimulate the postsynaptic end plate. This causes Ca^{2+} to be released into the cell sarcoplasm, which in turn binds to troponin, causing tropomysin to alter its position along the helix of actin filaments, resulting in muscle contraction (see Fig. 4A). The released ACH at the end plate immediately begins to be degraded by acetylcholinesterase and as soon as ACH concentration falls below threshold levels the entire process ceases. Ionic calcium also plays an important role since muscle relaxation results from the return of Ca^{2+} to the lateral sacs of the sarcoplasmic reticulum through the restoration of ATP by ADP. Through this process the nervous system both initiates and coordinates muscle activity. The electric activity in a muscle can be recorded through the help of electromyography by attaching suitable skin electrodes immediately over, or needles into, the muscles to be studied. Tracings using this approach have shown that electrical activity increases with the level of force developed.

V. EFFECTS OF EXERCISE ON MUSCLES

Skeletal muscle is adaptable and changes readily according to the activity patterns of daily life. Many tasks require strength, or isometric (static), effort, whereas others require endurance, or dynamic (isotonic), performance. When muscles are exercised regularly, training or physical conditioning results and affects the body as a whole. Isometric exercise occurs when muscle develops tension against resistance. During this form of exercise, the muscle is not allowed to extend, but remains in a state of heightened tension without visible evidence of useful work. This effect is experienced by the outstretched arm when supporting a weight or by the body's antigravity muscles when standing in place or maintaining posture. Repeated regular use of isometric exercise causes morphological and metabolic changes that increase muscle mass along with strength. Both an increase in fiber size (hypertrophy) and number (hyperplasia) are thought to occur, although the latter concept is still controversial (Brooks and Fahey, 1984). Physiologically, there is an increase in blood pressure and stroke volume, with little benefit to the cardiovascular system, since the resultant pressure load is without change in cardiac output. Because of the high sustained tension developed within the muscle, blood flow usually ceases despite the elevated mean blood pressure, so that this type of exercise may occur irrespective of oxygen supply.

Dynamic exercise, on the other hand, involves rhythmic movement of large

TABLE I

Consequences of Aerobic Training in Humans and Animals

Decreased resting heart rate
 at any given work level
  decreased heart rate
  decreased blood pressure
  more rapid recovery
  increased peripheral blood flow
Increased work capacity
Increased $\dot{V}_{O_2 max}$
Improved skeletal muscle metabolism
Increased cardiac mechanical performance and metabolism
Serum changes
 increased high-density lipoproteins
 decreased serum triglycerides
 decreased fasting blood sugar
 increased fibrinolysis

muscle groups and increases cardiac output, ventilation, and oxygen consumption. Such exercise is termed "aerobic," since it must be performed with sufficient oxygen present. Regular use of aerobic exercise results in significant general hemodynamic changes, as well as biochemical and morphological changes, in both cardiac and skeletal muscles.

In any type of aerobic training program, the final results depend on multiple factors including the initial level of fitness; physical endowment; and age, sex, and the presence or absence of disease. Physiological changes associated with such training are discussed in Chapter 7 and are summarized in Table I. Central hemodynamic adaptations (enhanced cardiac function and cardiac output) are observed in younger people (40 years and younger), whereas peripheral changes predominate in older individuals.

VI. DISUSE ATROPHY OF MUSCLE

Lack of use of muscle (bed rest, inactivity, or limb or body casts) and loss of innervation (disease or nerve severance) lead to a loss of muscle mass. As little as 1–2 months of disuse can lead to muscles that have atrophied to half normal size. If the individual is reactivated within that length of time, full function usually returns. After 4 months, however, a significant number of fibers usually degenerate, and full recovery is not possible. Re-innervation after 2 years rarely results in any function at all, because of replacement of muscle fibers by fat and fibrous tissue. During bed rest, the antigravity muscles of the legs are most affected, as is clinically evident by decreases in limb circumference, by urinary secretion of muscle breakdown products, and by reduced performance in tests of muscle strength. Deitrick and associates (1948) in a study of healthy individuals placed in body casts for 6–8 weeks found that the muscles associated with locomotion suffered the greatest decrease in strength. The biceps lost 6.6% of strength (measured by ergometer), the shoulder and arm muscles combined lost 8.7% (measured by straight-arm pull), the anterior tibial muscle group lost 13.3%, and the gastrocnemius-soleus group lost 20.8%. Strength of hand grip, back extensors, and abdominal muscles, on the other hand, did not decline. These changes probably represented an actual loss in muscle mass, since thigh circumference decreased by 3.5% and calf circumference by 5.6%, and were equivalent to decreases in cross-sectional areas of 5–10% in the thigh and 9–12% in the calf. Similar changes occurred during the Skylab flight (Thornton, 1981).

Histological studies using electron microscopy during the early phases of inactivity or immobilization show definite evidence of fiber disolution consistent with a catabolic process. As this process continues, areolar tissue appears between the fibers and begins to break up the muscle bundles. Areas of fibrosis are clearly evident by the sixth week of disuse and continue to proliferate throughout the remainder of the period of immobilization.

Muscle tissue, like other tissues, is maintained by balancing the rate of protein biosynthesis and of protein degradation. How the functional demands of muscle load and activity regulate this process is as yet unclear. It appears that immobilization, which decreases both activity and muscle load, also decreases contractile protein synthesis while increasing protein degradation, allowing the atrophic process to take place.

Several factors have now been clearly shown to be associated with the anabolic process. Principal among these are specific tissue growth factors that regulate required protein synthesis, and their lack, or altered presence, may either lead to, or contribute to, loss of muscle mass. These agents include insulin itself or in combination with growth hormone, fibroblast growth factor, and epidermal growth factor. Moreover, protein synthesis has long been known to be accelerated by such anabolic steroids as naturally occurring testosterone or its synthetic analog, methyl trienolone. But the manner of action of any of these factors on striated muscle is still poorly understood. The most reasonable hypothesis is that the aforementioned hormones and factors bind with the muscle cell sarcolemma, or specific receptors on the sarcolemma, or mitochondria to promote increased protein turnover. Muscle catabolism can be initiated by a decrease in number of receptors or a lack of affinity between growth factors, with the result that protein biosynthesis is reduced and intracellular proteolosis is increased. Active muscle apparently contains higher concentrations of endogenous metabolic stimulators of protein synthesis that are released into the muscle cell. They may accelerate protein synthesis in neighboring cells and lead to hypertrophy itself (S. Ellis, 1982).

Innervation of muscle is clearly important, for when nerve traffic is interrupted or decreased, muscle protein content rapidly decreases. Because of this observation, it has been suggested that a neurotropic factor or other chemical substances are released from the muscle at the myoneural junction (see Fig. 4), resulting in an anabolic stimulus that maintains muscle mass and regresses during immobilization. This hypothesis, however, is highly controversial. Doubt that a neurotropic factor is the sole explanation is based on findings that inactivity itself may generate an inhibitor of protein biosynthesis. This theory resulted from investigations of muscle regression induced by the use of thyroxine, in which an inhibitor of protein synthesis has been found in the postribosomal supernatant fraction. Inhibitors of this kind have also been found in serum and in culture medium studies using fibroblasts.

At present, the loss of contractile protein during inactivity is thought to be initiated by the release of Ca^{2+}-activated proteases from muscle. This protease has recently been shown to be located at the bands in the muscle fiber and is the point where the earliest indications of muscle atrophy appear. The process has recently been shown to be inhibited by the nontoxic microbial peptide, leupetin, with a marked decrease in muscle protein breakdown. Although Ca^{2+}-activated protease may be necessary to initiate the breakdown of contractile protein, lyosomes and their proteases also appear to play a further role in destroying this substrate. A great deal of further work is still needed to identify all the factors associated with these changes.

VII. EFFECTS OF INACTIVITY

Numerous bed rest studies have included evaluations of muscle atrophy with immobilization. Balaya and co-workers (1975) studied electromyographic changes in 12 subjects during a 30-day bed rest period in slightly head-up ($+6°$) and slightly head-down ($-6°$) body positions. Measurements were made in the flexors and extensors of the hands and feet. Curtailed muscular electrical activity occurred about halfway into the study and was associated with significant 1.5–2.5 cm) decreases in thigh circumference in all subjects. These changes which are indicative of lost muscle mass and function, were reversible with periodic isometric exercises in the post-bed-rest period.

J. P. Ellis and co-workers (1974) studied the effects of 2 weeks of horizontal bed rest on forearm amino acid metabolism in five subjects aged 18–20 years. No sigificant changes were found over the course of the study in either the uptake or the release of 19 amino acids. Of interest were findings of higher serum alanine levels in both arterial and venous blood after bed rest. It was hypothesized that this was the result of muscle breakdown in the more gravity-dependent muscles in other parts of the body.

Body casting in various animal models has been used more extensively to study muscle changes than it has been in humans (Wunder et al., 1968). Studies with casted animals have shown significant loss of muscle mass and atrophic change with full or partial body immobilization. Rats with hind legs casted and muscles fixed at lengths less than resting have shown significant muscle atrophy within 4–6 days (Booth, 1977). Measurements made under these conditions of fraction rate of protein synthesis have shown significant decreases within 6 hr of immobilization (Booth and Seider, 1979). Finally, comparative studies by Szilagyi and associates (1980) demonstrated that this type of immobilization closely simulates changes in muscles of rats flown aboard the 18.5-day Cosmos spaceflights.

When used in humans, body casting has also resulted in significant changes in muscle structure. The first classic study using body casting in humans was reported by Cuthbertson (1929) and involved normal male and female subjects of varying ages placed in single leg casts for periods of 10–14 days. Increases in urinary excretion of calcium, nitrogen, phosphorus, and sulfur were documented, confirming loss of bone and skeletal muscle mass. Later, Deitrick and associates (1948) observed minor changes in nitrogen excretion in normal, healthy males immobilized in plaster casts extending from the umbilicus to the toes over a period of 6–8 weeks. Creatinine excretion also increased and was associated with decreased circumferences of the lower limbs and decreased muscle strength. Other observers have noted decreases in muscle mass and power in the upper extremities with casting (Hettinger and Muller, 1953; Hislop 1963). Haggmark and Eriksson (1979) also found morphologic changes. Type I fibers (slow twitch) showed significant atrophy in eight patients following limb casting after surgery, as well as decreases in muscle oxidative enzymes. In a 22–year-old cross-country skier, casted for repair of a torn knee ligament, Eriksson (1980) reported a decrease in Type I fiber content from 80% at the time of surgery to 57% after 1 month in the cast. Following retraining, the Type I population returned to 84% in similar studies of patients casted for fracture of a single leg. Sargeant and associates (1977) observed an average decrease of 12% in volume of the casted leg and 42% in the cross-sectional area of biopsied muscle fibers (both Type I and Type II fibers). These subjects further showed a 16.9% decrease in $\dot{V}_{O_2 \text{ max}}$ during one-legged bicycle pedaling. Similar changes in maximal oxygen uptake had been shown previously by Davies and Sargeant (1975a,b), who reported a 12% decrease in $\dot{V}_{O_2 \text{ max}}$ during one-legged pedaling in subjects immobilized for 15 weeks in long-leg plaster casts because of fractures. Volume of the affected limb also was seen to decrease by 12%.

VIII. SPACEFLIGHT FINDINGS

Exposure to spaceflight has also consistently resulted in loss of muscle mass in both astronauts and cosmonauts. On return to earth, space crews generally have had difficulty in walking and maintaining an upright posture. All cosmonauts after very long stays in space (185–237 days) have even had to be carried off their returning vehicles.

Skylab flights (28-, 56-, and 84-day missions) included physiological measurements before, during, and after flight. Findings have indicated significant losses of muscle mass, as documented by an increased urinary excretion of nitrogen and phosphorus (Whedon *et al.*, 1977). These changes have also been accompanied by significant decreases in lower leg circumferences (Johnson *et al.*, 1977). Tests for muscle strength and endurance also decreased after the Skylab flights (Thornton and Rummel, 1977). Leg extensor strength decreased on the average by 10–16% and was associated with a leg volume decrease of 7–11%. These findings are important, because the tests were conducted 5 days after flight and each successive mission included increased use of in-flight exercise. On this basis, it is interesting to note that arm extensor strength decreased by 10% in the Skylab 2 and 4 astronauts, whereas those from Skylab 3 showed no change. Arm flexor strength decreased by about 8% in the Skylab 2 crew but did not change in the crewmen of the other missions. Finally, leg extensor strength decreased by about 20% in the first two flights but remained unchanged in the last.

A 1–2% loss of body mass has been found regularly during the first few days of flight but is usually replaced rapidly after recovery (Thornton and Ord, 1976). These changes probably represent fluid losses resulting from the headward shift of body fluids and blood triggered by weightlessness. Loss of muscle mass had followed when a weightless or inactive state is continued. These changes have been confirmed by other measurements made after flight, which include total body water, total body potassium (K^{40}), plasma volume, total body volume by stereophotography, and metabolic balance. Plasma volume decreases induced by weightlessness were 3.3%, 13.1%, and 15.9%, respectively, in the successive Skylab flights. Slightly more than half of the weight loss observed during these missions resulted from losses of lean body mass, with protein losses stemming primarily from disuse atrophy of antigravity muscles (Leach *et al.*, 1979).

Since the Skylab missions, Soviet cosmonauts have made much longer flights. Following a 185-day flight both cosmonauts showed decreased locomotor function and an inability to coordinate movements (Gazenko *et al.*, 1981). Immediate postflight studies showed decreases in volume and circumference of the lower extremities. Biopsies showed loss of muscle mass, which was masked by an increase in adipose tissue. Examination on the fourth day of recovery revealed atrophic changes in the long muscles of the back, as well as in the chest and the abdominal muscles. In these areas, muscle size, velocity, strength, and tone had all decreased.

Soviet investigators have also flown a number of animal flights using rats in their Cosmos series (605, 690, 782, 936, and 1129). A significant decrease in the mass of antigravity muscles was seen (Gazenko *et al.*, 1980). Decreases were also seen in the content of myofibrillar and sarcoplasmic proteins in the soleus, and glycerated myofibers showed reduced contraction strength and work capacity (Oganov, 1977). Finally, morphometric studies showed a reduction in the total area of fibers and specific losses of Type I (slow-twitch) fibers (Chui and Castleman, 1980).

IX. MANAGEMENT OF DISUSE ATROPHY

Our present sedentary life as well as injuries, immobilization, disease, and aging lead to a condition of weakened muscles. As already mentioned, disuse atrophy can be counteracted by therapeutic exercise, which restores muscle mass and strength. Both isometric and isotonic exercise are useful for these purposes. In most cases, the mere return to normal activity is sufficient to restore muscle mass and strength after immobilization, but the process may be very slow. Stressing muscles beyond normal levels of activity will accelerate the process of return. In general, the goal of all physical fitness or conditioning programs is to gradually increase the load and work requirement, or both, until the desired level of activity is reached. A gradual increase in load ensures minimal incidence of injury to muscles and joints and ensures repeated stimulation of affected muscles. Specific programs may be complicated, and a number of investigators have attempted to simplify it. Muller (1970) reported that 6 sec of contraction at near-maximal load once a day created a significant increase in strength. Rose and associates (1957) developed a brief maximal exercise system in which a maximal load was lifted once a day from 90° of knee flexion to full extension and was held in an isometric contraction for 5 sec. The load was increased by a small amount each day. Once the desired strength was achieved, it could be maintained by performing the exercise only once a week.

When joints are painful and inflamed, isometric exercise may be the preferred procedure to avoid further stress (Steinberg, 1980). In this type of exercise, the joint is not moved and the muscle works a static contraction. Individuals can also will a partially denervated muscle to contract, but the contraction force is much less than normal function. In the case of such muscles, it is not yet clear whether exercise strengthens the muscle or further fatigues it. Children with birth injuries or poliomyelitis have been studied to evaluate the effects of training on unused paretic muscles (Muller and Beckman, 1966). The investigators found that 50% of the inactive muscles could be trained to function at about half of normal strength. They concluded that most of the muscle fibers (80%) were not damaged but had simply become inactive from disuse. Children with birth injuries showed better function than those with poliomyeltis effects. Others have reported, however, that partially denervated muscle in poliomyelitis and traumatic quadriplegia cases were weakened by overuse (Bennett and Knowlton, 1958). These investigators used the term "overwork weakness" for the condition and concluded that it could result in permanent damage. Normally, an individual does not overwork his or her muscles, because metabolic end products accumulate and create feelings of fatigue and discomfort, so that the individual ceases the effort. But this does not occur with damaged muscle, because there is a greater ratio of blood supply to muscle fiber, and metabolic end products do not accumulate sufficiently to establish a limit of activity. Consequently, when muscle weakness results from injury or neurologic diseases, those dealing with muscle disuse atrophy should assess the potential effects of any exercise program that involves repeated muscle testing. If a muscle is weaker after exercise, it has been stressed excessively, particularly if the weakness is still apparent on the following day. Clinicians have observed that neuropathic patients are often stronger on Mondays after they have had 2 days of relief from therapeutic exercise (Steinberg, 1980). Therefore, such findings point to the need for careful attention in the administration of any exercise program that takes into account the characteristics of the disease itself, as well as the condition of each individual's musculature following immobilization.

Although totally denervated muscles will no longer respond to signals from the central nervous system, they do respond to electrical stimulation. The contractions seen with electrical stimulation may limit or prevent muscle atrophy with immobilization. This approach is used principally where there are motor neuron lesions resulting from nerve injury, toxins, or disease. Frequent electrical stimulation, for example, is used in poliomyelitis patients to maintain muscle function, since anterior horn cells may recover or since sprouting of nerve endings may occur over time. The procedure, however, does not help myopathic muscle weakness or lesions of the myoneural junction but may be useful in lesions of the central nervous system. In the latter case, the motor units may be intact but can receive no messages from the central nervous system and as a result may atrophy. Electrical stimulation forces the muscles to contract when they cannot do so through natural means, thus reducing the potential for muscle degeneration.

The electrial impulses cause a contraction only when the electrical flow begins and ends. Therefore, the current used must either be alternating or consist of rapid, unidirectional short impulses. Currents whose pulses reach a peak almost instantaneously and persist for about 1 msec are most often used for therapeutic purposes. Current strength has varied from 1 to 30 mA. Electrical stimuli have a greater effect on nerve fibers than on muscle fibers. Consequently, intact nerves will respond to an electrical impulse and will create muscle contractions at a much lower current strength than would be required for stimulating muscle fibers. Intact nerve fibers react to electrical impulses lasting 1–100 msec and having the same threshold of current intensity, or rheobase. For shorter duration times (1 msec), a higher current intensity is needed to make the muscle contract.

Current intensity must be increased if still shorter durations are desired, and the current intensity can then become painful. Figure 5, taken from Steinberg (1980), demonstrates the relationship between current intensity and impulse duration to create muscle contraction in both normal and denervated nerves. This "strength–duration curve" is used diagnostically to test for denervation.

Damaged nerves, on the other hand, do not react to electrical impulses. Then the muscle fibers must be directly stimulated, but at a much higher current intensity and pulse duration, about 100 times greater than what the intact nerve would require. Even at these higher thresholds, muscle fiber reacts to electrical stimulation with a less forceful, slower contraction.

During electrostimulation, the impulse frequency must be rapid enough so that the muscle cannot relax completely between impulses. The more rapid the impulses, the less time between contractions. When the nerve is intact, impulses spaced at 20 per second permit no relaxation and result in a tetanic contraction that continues until the current is stopped. The same applies to muscle fibers, but the contractions again are less forceful and slower.

With electrostimulation, denervated muscle must be treated two to three times each day to prevent atrophy and degeneration. This can be accomplished in the hospital, in an out-patient setting, or at home. Many patients have been taught to follow required clinical procedures, with small battery-powered stimulators that generate a direct current and can be controlled manually by a switch. In addition, there are more advanced systems to accomplish these ends, as shown in Fig. 6. Here, computer control is used for alteration of associated stimulation units.

Soviet investigators have conducted a number of in-depth studies to determine the benefits of muscle electrostimulation in bed rested subjects. Methods for electrostimulation of the lower portions of the body have varied considerably with regard to the number of electrodes used (10–20), amount of current, and

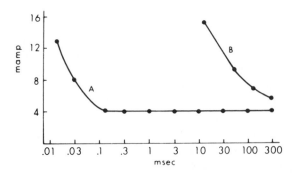

Fig. 5.   Strength–duration curve. (A) Normal muscle. (B) Denervated muscle. [From Steinberg, 1980, with permission.]

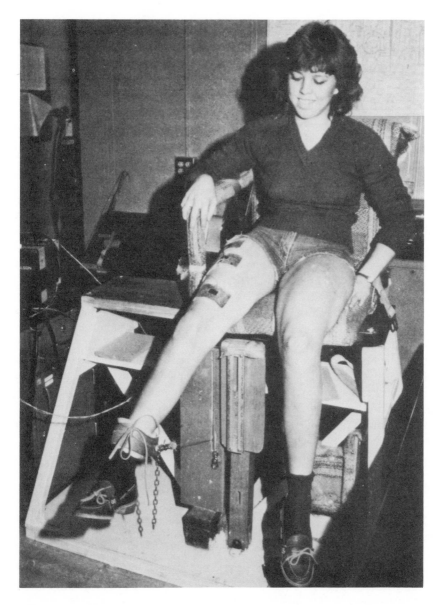

Fig. 6. Computer-controlled, progressive-resistance electrical stimulation of paralyzed muscle.
[From Brooks and Fahey, 1984, with permission.]

duration of use (from 15 min to several hours a day) (Dukhin and Zukovs'kiy, 1979; Georgiyevskiy *et al.*, 1979; Kakurin and Gregor'yev, 1981; Mikhaylov and Georgiyevskiy, 1976). Most often, a portable, multichannel electrostimulator has been used. The technique has been applied alone or in conjunction with other methods, such as physical training. Georgiyevskiy and associates (1979), after a 45-day study of 12 healthy male subjects (22–35 years old) immobilized in the head-down position (−6.5°), concluded that myoelectrostimulation could effectively prevent atrophy of skeletal muscle. They based their conclusion on post-bed-rest muscle biopsies that showed muscle atrophy in untreated control subjects but not in the electrostimulated group. Soviet investigators have also found as much as a 40% loss in exercise capacity ($\dot{V}_{O_2 \, max}$) after bed rest in untreated subjects, which could either be improved or prevented by electrostimulation, yet there was failure to reverse or to prevent evidence of orthostatic intolerance with head-up tilt or lower-body negative pressure (Kakurin *et al.*, 1976, 1980; Kakurin and Gregor'yev, 1981; Yegorov, *et al.* 1970). Following a 30-day bed rest study, Chekirda and associates (1974) compared the gait responses of subjects who were trained in walking and running in the supine position with those of subjects who were not trained but were administered electrostimulation of the muscles. Since the exercise group performed better in gait tests than the electrostimulated group did, the investigators concluded that physical training was more beneficial than electrostimualtion of muscles. Cherepakhin and associates (1977), on the other hand, concluded from the findings of a 49-day bed rest study, that a combination of physical training and muscle electrostimulation had a positive preventive effect on the muscle problems associated with immobilization and weightlessness. Electrostimulation of muscle continues to be a widely used and respected method among Soviet investigators as a countermeasure against atrophic loss during immobilization and inactivity.

X. DISEASES OF MUSCLE INACTIVITY

Although prolonged inactivity and immobilization are detrimental to the circulatory system and heart function, as well as perhaps to metabolism and digestion, they are particularly deleterious to skeletal muscle. When muscles are not used sufficiently, they weaken and deteriorate, and as a result cause a number of orthopedic health problems. As early as 1868, Taylor (Goldswaith *et al.*, 1945) recognized the impact of lack of muscle function on back pain, poor posture, and other orthopedic diseases. Findings by Goldswaith and associates (1945) supported Taylor's conclusions, but these results on the effects of muscle disuse have only recently become generally accepted. One of the most common of today's medical ailments, low back pain, can be traced to a process in which muscles become too weak to support the skeletal structure properly. Due to our sedentary life style, as well as to injuries, immobilization, and disease, our muscles do less work and weaken. Sedentary individuals over time use less adequately the large muscle groups that maintain the body posture. This is also particularly the case for subjects with neuromuscular diseases and fractures and for those immobilized for any reason. Muscle tone in general becomes slack; this is most notable with the anterior abdominal muscles, which sag when the body becomes erect. Muscles of the lower extremities that are not exercised regularly lose mass and become spindly. The loss of muscle function and strength is especially apparent in the elderly, who become less and less active either through inability or disinterest. Danger from falling and bone fracture are especially great since there may not only be loss of muscle mass from inactivity, but bone mass and strength from osteoporosis. Regular exercise can reduce these health problems in both the young and the elderly (Bortz, 1982). Even if there has been some loss through disuse, a gradually increasing muscle loading by mild exercise can improve the capability of disused muscles to better support the skeletal structure.

XI. CONCLUSIONS

Various levels of inactivity and immobilization can cause deterioration of muscle structure, function, and strength. These results have been seen not only in clinical and experimental settings, but also in weightlessness during spaceflight. Muscle disuse atrophy can be reversed fairly rapidly after short periods of immobilization, but the longer the atrophy exists, the longer the time needed to reverse the problem. After too long a period, muscle disintegration from disuse is no longer reversible. Exercise and electrostimulation are used to return ailing muscle to a healthy condition, but are not always successful. Additional countermeasures are needed, particularly in view of plans for very long term stays in space. Before they can be achieved, however, we will need a much broader knowledge of the biochemical causes of muscle disuse atrophy. Therefore, considerable research remains to be done.

REFERENCES

Balaya, N.A., Amirov, R.Z., Shaposhnikov, Ye.A., Lebedeva, I.P., and Sologub, B.S. (1975). *Voprosy Kurortologii, Fizioterappi i Lechebnoy Fizicheskoy Kul'tury* **40**, 238-241.
Bennett, R.L., and Knowlton, G.C. (1958). *Clin. Orthop.* **12**, 22.
Booth, F.W. (1977). *J. Appl. Physiol.* **43**, 656-661.
Booth, F.W. and Seider M.J. (1979). *J. Appl. Physiol.* **47**, 974-977.
Bortz, W.M. (1982). *JAMA* **248(10)**, 1203-1208.
Brooks, G.A. and Fahey, T.D. (1984). "Exercise Physiology." John Wiley & Sons, New York.
Browse, N.L. (1965). "The Physiology and Pathology of Bed Rest." Charles C. Thomas, Springfield, Illinois.

Chekirda, I.F., Yeremin, A.V., Stepantsov, V.I., and Borisenko, I.P. (1974). *Space Biol. and Aerosp. Med.* **8(4)**, 69-74.

Cherapakhin, M.A., Kakurin, L.I., Il'ina-Kakuyeva, Ye.I., and Fedorenko, G.T. (1977). *Space Biol. and Aerosp. Med.* **11(2)**, 87-91.

Chui, L.A. and Castleman, K.R. (1980). *The Physiologist* **23**, S83-S86.

Cuthbertson, D.P. (1929). *Biochem. J.* **23**, 1328-1345.

Davies, C.T.M. and Sargeant, A.J. (1975a). *Scand. J. Rehab. Med.* **7**, 45-50.

Davies, C.T.M. and Sargeant, A.J. (1975b). *Clin. Sci. Molec. Med.* **48**, 107-114.

Deitrick, J.E., Whedon, G.D., and Shorr, E. (1948). *Am. J. Med.* **4**, 3-36.

Dukhin, Ye.O. and Zukovs'kiy, L.Y. (1979). *Visnykh Akademii nauk USSR* **10**, 48-54. (NASA TM-76318).

Ellis, J.P., Lecocq, F.R., Garcia, J.B., and Lipman, R.L. (1974). *Aerosp. Med.* **45**, 15-18.

Ellis, S. (1982). *Disuse Muscle Atrophy.* NASA Research and Technology Resume. National Aeronautics and Space Administration, Ames Research Center, Moffett Field, California.

Eriksson, E. (1980). *Contemp. Orthop.* 2, 228-232.

Gazenko, O.G., Genin, A.M., Ilyin, E.A., Oganov, V.S., and Serova, L.V. (1980). *The Physiologist* **23**, S11-S15.

Gazenko, O.G., Genin, A.M., and Yegorov, A.D. (1981). *Acta Astronautica* **8**, 907-917.

Georgiyevskiy, V.S., Il'inskaya, Ye.A., Mayeyev, V.I., Mikhaylov, V.M., and Pervushin, V.I. (1979). *Space Biol. and Aerosp. Med.* **13(6)**, 52-57.

Goldswaith, J.E., Brown, L.T., and Swaim, L.T. (1945). "Essentials of Body Mechanics in Health and Diseases. J.B. Lippincott Co., Philadelphia, Pennsylvania.

Haggmark, T. and Eriksson, E. (1979). *Am. J. Sports Med.* **7**, 48-56.

Hettinger, T. and Muller, E.A. (1953). *Arbeits Physiol. Angew. Entomol.* **15**, lll.

Hislop, H.J. (1963). *J. Am. Phys. Ther. Assoc.* **43**, 21.

Johnson, R.S., Hoffler, G.W., Nicogossian, A.E., Bergman, S.A., Jr., and Jackson, M.M. (1977). *In* "Biomedical Results from Skylab" NASA SP-377, pp. 284-312. National Aeronautics and Space Administration, Washington, DC.

Kakurin, L.I. and Grigor'yev, A.I. (1981). *In* "Proceedings of the 12th US/USSR Joint Working Group on Space Biology and Medicine."

Kakurin, L.I., Yegorov, B.B., Il'ina, Ye.I., and Cherepakhin, M.A. (1976). *In* 'International Symposium on Basic Environmental Problems, Man in Space" (Graybiel, ed.), pp. 241-247, Pergamon Press, Oxford.

Kakurin, L.I., Grigor'yev, A.I., Mikhaylov, V.M., and Tishler, V.A. (1980). *In* 'Proceedings of the llth US/USSR Joint Working Group on Space Biology and Medicine," pp. 1-57. (NASA TM-76465).

Leach, C.S., Leonard, J.I., and Rambaut, P.C. (1979). *The Physiologist* **22**, S61-S62.

Mikhaylov, V.M. and Georgiyevskiy, V.S. (1976). *Space Biol. and Aerosp. Med.* **10(6)**, 56-63.

Muller, E.A. (1970). *Arch. Phys. Med. Rehab.* **51**, 449.

Muller, E.A. and Beckman, H. (1966). *Z. Orthop.* **102**, 139.

Oganov, V.G. (1977). "Proceedings of the 8th US/USSR Joint Conference on Space Biology and Medicine." NASA TM-75073, pp. 1-40. National Aeronautics and Space Administration, Washington, DC.

Rose, D.L., Radzyminski, S.F., and Beatty, R.R. (1957). *Arch. Phys. Med. Rehab.* **38**, 157.

Sargeant, A.J., Davies, C.T.M., Edwards, R.H.T., Maunder, C., and Young, A. (1977). *Clin. Sci. and Molec. Med.* **52**, 337-342.

Steinberg, F.U. (1980). "The Immobilized Patient." Plenum Medical Book Company, New York.

Szilagyi, A., Szoor, E., Takacs, O., Rapcsak, M., Oganov, V.S., Skuratove, S.A., Oganesyan, S.S., Murashko, L.M., and Eloyan, M.A. (1980). *The Physiologist* **23**, S67-S70.

Thornton, W.E. (1981). *In* "Spaceflight Deconditioning and Physical Fitness" (J.F. Parker, Jr., C.S. Lewis, and D.G. Christensen, eds.). National Aeronautics and Space Administration, Washington, DC, pp. 13-81.

Thornton, W.E. and Ord, J. (1976). *In* "Biomedical Results from Skylab" NASA SP-377, pp. 175-182. National Aeronautics and Space Administration, Washington, DC.

Thornton, W.E. and Rummel, J.A. (1977). *In* "Biomedical Results from Skylab." NASA SP-377, pp. 191-197. National Aeronautics and Space Administration, Washington, DC.

Vander, A.J., Sherman, J.H., and Luciano, D.S. (1970). "Human Physiology: The Mechanisms of Body Function." McGraw-Hill, Inc., New York.

Whedon, G.D., Lutwak, L., Rambaut, P.C., Whittle, M.W., Smith, M.C., Reid, J., Leach, C., Stadler, C.R., and Sanford, D.D. (1977). *In* "Biomedical Results from Skylab" NASA SP-377, pp. 164-174. National Aeronautics and Space Administration, Washington, DC.

Wunder, C., Duling, C.B., and Bengele, H. (1968). *In* "Hypodynamics and Hypogravics" (M. McCally, ed.), pp. 71-108. Academic Press, New York.

Yegorov, B.B., Georgiyevskiy, V.S., Mikhaylov, V.M., Kil, V.I., Semeniutin, I,P., Kazmirov, E.K., Davidenko, Yu.V., and Fat'ianova, L.I. (1970). *Space Biol. and Med.* **3(6)**, 96-101.

5

Metabolic and Endocrine Changes

JOAN VERNIKOS
Cardiovascular Research Office
Biomedical Research Division
National Aeronautics and Space Administration
Ames Research Center
Moffett Field, California 94035

I. METABOLIC CHANGES WITH INACTIVITY

A. General

The most basic changes occurring with reduced activity would be expected to involve the basal metabolic rate (BMR), caloric requirements, and dietary intake. In fact, the BMR drops slightly during prolonged immobilization. In a study of healthy young men immobilized from the umbilicus to the toes in body casts for 6–7 weeks, Deitrick and associates (1948) observed a decline of 6.9% in the BMR. This change appears within 20–24 hr and remains stable thereafter, despite the length of inactivity. The BMR usually returns to normal 3 weeks after the subject returns to active life (Browse, 1965).

Most individuals subjected to prolonged bed rest lose weight, regardless of whether they are healthy volunteers participating in a research project or patients bed rested for health problems, and whether they are on controlled diets or are allowed free choice of food (Greenleaf *et al.*, 1977). The weight loss results from a combination of factors: the known diuresis occurring with the onset of bed rest, the subsequent shift from lean body mass to body fat, and finally, a decrease in caloric need due to inactivity.

Any evaluation of changes in body weight during bed rest must take into account the energy balance of the subjects and the changes in lean body mass versus fat. In a comparison of men and women after 2–3 weeks of bed rest, Pace and associates (1976) reported that both sexes lost body mass while body fat remained unchanged or increased slightly. Greenleaf and associates (1977)

99

studied the effects of exercise on the body composition of 7 healthy young men (19–21 years) performing isotonic and isometric exercise during 2-week bed rest periods. Body weight decreased slightly (−0.43 kg) during a period of no exercise but also decreased significantly with both isotonic (−1.77 kg) and isometric (−0.91 kg) exercise. About one-third of the weight lost with isotonic exercise was caused by a loss of body fat (−0.69 kg), and the remainder by a loss of lean body mass (−0.98 kg). The loss of body fat appeared proportional to the decrease in metabolic rate, whereas the loss of lean body mass was independent of metabolic rate and resulted from the horizontal body position.

Caloric requirements during the inactivity of bed rest are increased by physical activity, tissue injury, fever, or metabolic derangements. Hodges (1980) has suggested that in most cases, 30–40 kcal per kilogram of body weight per day will meet the dietary needs of immobilized individuals. For an active person, the diet should contain about 50% complex carbohydrates, 40% fat, and 10% protein. Most inactive or immobilized individuals seem to experience a loss of appetite, particularly for proteins, which are important at this time to offset protein catabolism, in which muscle mass is metabolized while body fat is stored. For the immobilized indiviudal then, it has been traditionally considered desirable from a metabolic point of view to reduce the dietary content of carbohydrates and fats and to increase the intake of protein. Hodges (1980) has proposed that the diet during immobilization should contain 40% complex carbohydrates, 28% fat, and 32% protein. Although the reduction in fats and the increase in protein intake seem logical, the reduction in carbohydrates may have undesirable consequences on the functional capacity of the sympathetic nervous system (Young et al., 1973), as will be discussed later.

It has been reported that a change of diet, lack of appetite, and other effects associated with bed rest may adversely affect immune mechanisms (Bistrian et al., 1975; Law et al., 1974; Munster, 1976) and that a mere 10 days of caloric restriction can significantly alter the metabolism of tryptophan and niacin in young men (Consolazio et al., 1972). Clinical evidence of these altered metabolic responses can occur within 2 weeks after immobilization and has been said to affect as many as 50% of immobilized hospital patients (Butterworth, 1974).

Thyroid hormone concentrations become less stable during prolonged bed rest (Vernikos-Danellis et al., 1974). In a study involving 8 healthy male subjects bed rested for 56 days, we found that diurnal rhythms of thyroid activity became distinctly unstable (Vernikos-Danellis et al., 1974). These alterations in thyroid activity may affect the changes in body composition seen with prolonged bed rest. In 115 spinal cord injury patients, Browse (1965) reported serum thyroxine, thyroid stimulating hormone, and T_3 uptake to be within normal limits. Serum T_3 levels, however, tended to be low in both paraplegic and quadriplegic patients despite the fact that they were clinically euthyroid and that other thyroid functions were normal.

In spite of general interest in the beneficial effects of increased activity on levels of triglycerides and cholesterol, the changes in these that may occur with prolonged immobilization have not received a great deal of attention. The Soviet investigators Vendt and associates (1979) studied subjects bed rested in the head-down position to determine quantitative and qualitative changes in sterols bound with plasma proteins and erythrocyte membranes. They found a phasic pattern of changes in high-density lipoproteins (HDL) and cholesterol, with low-density lipoproteins (LDL) generally changing in the opposite direction. Red blood cell membranes showed a change in strength and permeability and a gradual increase in cholesterol content. Cholesterol content of membranes reached a maximum on day 48, while LDL reached a minimum. After 14 days of recovery from bed rest, cholesterol in membranes returned to normal, but there was a significant increase in transport proteins. Thin-layer chromatography, gas–liquid chromatography, and ultraviolet and infrared spectroscopy showed that sterols—the breakdown products of cholesterol and adrenal compounds—increased at an early stage in bed rest, increased again in the latter stages, and did not return to normal throughout recovery.

The pituitary–adrenal system plays a key role in maintaining homeostasis. In the earlier bed rest studies, urinary excretion of 17-hydroxycorticosteroids (17-OHCS), or 17-ketosteroids, which are the breakdown products of cortisol, were used by a number of investigators to measure adrenal gland function over the course of bed rest, but the reported changes were inconsistent (Cardus et al., 1965; Deitrick et al., 1948; Katz et al., 1975; Sandler et al., 1978; White et al., 1966). On the other hand, these and more recent studies have shown increased urinary cortisol excretion during bed rest in males and an even greater increase during ambulatory recovery. These differences may reflect the problems of attempting to use urinary values of adrenal breakdown products as indicators of gland secretion. They also may reflect the response of individuals to the bed rest scenario, depending on whether participating in a bed rest study or lying in bed for several days is perceived by the individual as confining and associated with helplessness or as a time for relaxation. We have observed significant sex differences in the urinary excretion of cortisol in response to bed rest in directly comparable studies, which we believe are associated with the extent to which individuals perceive bed rest as a stressful situation (unpublished observations).

The pituitary–adrenal system is no exception to the apparent uncoupling between the driving system and target organ that has been observed in numerous other systems during bed rest and inactivity. Changes in plasma adrenocorticotropin (ACTH) and cortisol levels over a 56-day bed rest period are shown in Fig. 1. It is apparent that sensitivity of the adrenal to ACTH takes time to develop since decreased target organ responsiveness is not significant at 6 days of bed rest as shown by the plasma cortisol response to an infusion of ACTH in Fig. 2 (Dallman et al., 1984). However, a decreasing sensitivity of the adrenal to

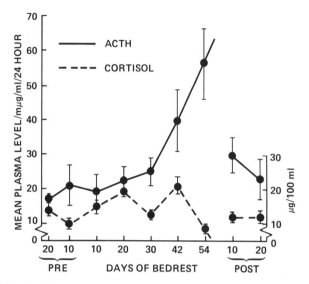

Fig. 1. Mean circulating cortisol and ACTH per 48-hr sampling period during 56 days of bed rest; vertical lines represent SE ($N = 5$).

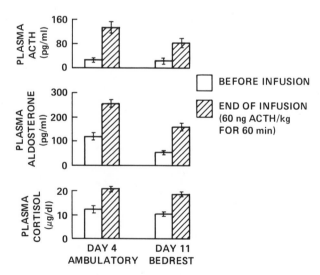

Fig. 2. Plasma ACTH, aldosterone and cortisol levels before and at the of ACTH infusion during the ambulatory control and head-down bed rest periods. Bars representing mean data from 8 subjects are accompanied by vertical lines indicating ±SEM.

ACTH may be apparent by 14 days. The adrenal response to centrifugation in female subjects 14 days after bed rest was unchanged in spite of greatly enhanced plasma ACTH responses to the same stimulus as shown in Fig. 3 (Vernikos-Danellis *et al.*, 1978). We have also observed this effect to occur normally as a function of age (see Fig. 4) suggesting that it may, indeed, be a consequence of the reduced activity with age rather than due to aging itself (Vernikos, 1984). In the 56-day bed rest study (Fig. 1), in which no provocative tests were used, target organ responsiveness was seen to decrease after 30 days; plasma cortisol levels were decreasing as circulating ACTH increased to three times that found in ambulatory subjects, and both returned to normal within 20 days of the subjects' becoming ambulatory again. Such findings are consistent with those of Kaplan and co-workers (1966) and Vallbona and co-workers (1966), who found similar results of decreased adrenal responsiveness in permanently inactive individuals. This has led Steinberg (1980) to suggest that such individuals would need support by exogenous adrenocorticoids to prevent severe hypotension when in an upright position.

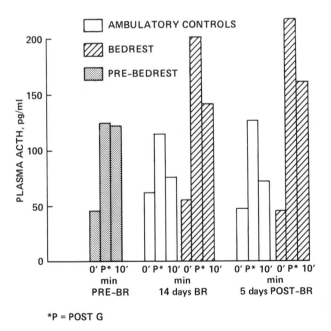

Fig. 3. Plasma ACTH responses to $+3$ G_z acceleration in normal female subjects before and after 14 days of bed rest or confinement.

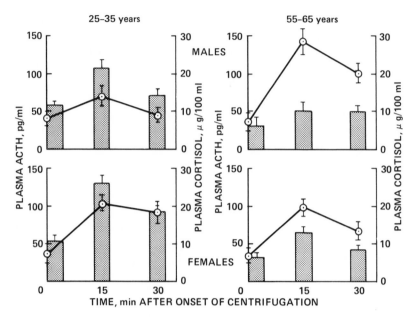

Fig. 4. Plasma ACTH (O—O) and cortisol (solid bars) responses to the stress of centrifugation in male and female subjects 25–35 and 55–65 years old. Each point represents the mean concentration ±SEM.

B. Salt and Water Metabolism

Weightlessness and bed rest alike produce a characteristic headward redistribution of blood volume. The shift of approximately 700 ml of blood toward the upper part of the body results in increased intrathoracic pressure and stimulation of cardiopulmonary receptors and in the stimulation of aortic and carotid baroreceptors. Such stimulation results in an inhibition of the sympathetic nervous system, of the renin–angiotensin–aldosterone system and of the antidiuretic hormone (ADH). In contrast, prostaglandin and a natriuretic factor would be increased. Diuresis and natriuresis (increased sodium excretion) and the resultant decrease in blood volume constitute the homeostatic response to the postural change. The overall endocrine and neurohumoral processes are schematically shown in Fig. 5.

Significant fluid shifts have been measured at various stages of bed rest: the first day, the first 2 weeks, and the first month. The first day, a diuresis occurs and plasma volume is reduced (Sandler, 1980; Taylor *et al.*, 1945). This is regularly associated with a natriuresis. The diuresis occurs regardless of the amount of total and voluntary intake of fluids. In a 120-day bed rest study, a periodicity of diuresis was detected. During the first 36 days, there was an average 315-ml diuresis; by day 53 the diuresis decreased; by day 83 it increased

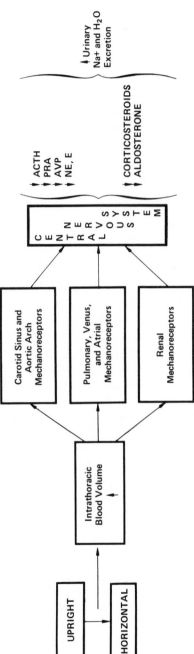

Fig. 5. Primary mechanisms in the regulation of plasma volume and electrolytes following postural change.

once more. The result was a predominantly negative water balance. The decrease on day 53 was preceded by a period of sodium retention that activated aldosterone and led to ADH suppression and the resultant recurrence of a diuresis. A study by Krotov *et al.* (1977) confirmed this rhythmicity in water volume regulation. Using tritiated water, altered elimination was shown, both during long-term bed rest and during the 2–3 weeks of the recovery period as well.

The headward shift in body fluids results in an expected reduction of the volume of the lower extremities. The volume of the thighs and upper legs decreases by 6%, the lower legs by 5%, and the circumference of the calves by 2.7% (Blomqvist *et al.*, 1980; Greenleaf, 1984; Hargens *et al.*, 1983). Using tracer and isotope markers and other techniques, it can be shown that of the 500–700 ml transferred cephalically, about half comes from the vascular spaces of the pelvis, thighs, and lower legs, with the rest coming from extravascular spaces including subcutaneous tissue, muscle, and a small amount from bone itself (Greenleaf, 1984). This fluid shift causes central venous pressure to increase, triggering mechanisms that bring about a diuresis. The loss of fluid by diuresis, which results in a negative water balance, contributes to the loss observed in total body water (intracellular and extracellular). Extracellular fluid is the primary contributor to the fluid loss in the early days of bed rest. By the second day of bed rest, total body water decreases by about 600 ml (Jacobson *et al.*, 1973). By the fourth day, extracellular volume has decreased by 400 ml, and plasma volume by 440 ml (Greenleaf and Koslowski, 1982). Twelve to 14 days of bed rest results in a decrease in total body water of 1,480 ml; plasma volume remains reduced by 500 ml, but extracellular volume returns to ambulatory control values. By this time, interstitial volume (extracellular minus plasma volume) has increased to 300–400 ml—an amount equal to the decrease in plasma volume (Greenleaf *et al.*, 1977; Jacobson *et al.*, 1973; Melada *et al.*, 1975). When total body water decreases while extracellular volume remains normal, it is obvious that the loss must come from intracellular fluids. Since extracellular volume returned to normal after 14 days of bed rest, Greenleaf (1984) has concluded that the body places greater importance on maintaining extracellular volume than on maintaining plasma volume and that interstitial volume may be the controlling factor.

Plasma volume for the normal male is usually 40 ml/kg body weight and is decreased by 5.3–9.7% after 24 hr of horizontal bed rest (Greenleaf, 1984). For a person weighing 80 kg, this represents losses ranging from 150 to 650 ml. The majority of plasma volume losses are seen within the first 24–72 hr, with only slow change thereafter. However, data on plasma volume changes in bed rest longer than 2 weeks are limited. The few lengthier studies have shown that plasma volume continues to decrease. Compared with ambulatory pre-bed-rest measurements, plasma volume was decreased by 15% after 20 days of bed rest, by 18% after 70 days, and by 30% after 175 days (Donaldson *et al.*, 1969). After

25 days of bed rest, total body water decreased by 3.4%, with about one-third of that stemming from plasma volume and the remainder from intracellular volume. Red blood cell mass decreases more slowly than plasma volume, dropping by 100 ml after 14 days of bed rest and by 300–550 ml after 30 days (Kimzey et al., 1979; Miller et al., 1965). Some investigators have suggested that the hypovolemia that occurs with bed rest can be stabilized by the performance of isotonic exercise (Greenleaf et al., 1977). Data from bed rest and spaceflight have failed to bear this out (see Chapter 2).

When the body stands up, plasma volume decreases by about 5% within 30 min and plasma renin-angiotensin (PRA), aldosterone, vasopressin (AVP), ACTH, and cortisol increase, whereas the reverse is true on lying down. Within 5 min of assuming the head-down ($-6°$) position, there is a significant decrease in heart rate that is sustained for 2 hr before gradually returning toward normal during the next 6 hr. No changes occur in indirect systolic or diastolic arterial blood pressure during the first 24 hr after lying down. However, there are prompt and sustained decreases in AVP, PRA, and aldosterone over the first 8 hr. Circulating PRA reaches a nadir by 2 hr that is sustained at 4 and 8 hr but returns to normal by 24 hr. In contrast, plasma aldosterone, which is believed to be regulated primarily by PRA, has its nadir at 4 hr, and although the values increase gradually during the next 20 hr, aldosterone levels are still depressed at 24 hr (Dallman et al., 1984).

The rapid inhibition in the levels of hormones that regulate salt and water metabolism on lying down is responsible for the changes in renal fluid and sodium (Na) excretion during the first days of bed rest. There is a diuresis and a marked loss of fluid and sodium (negative fluid and sodium balance) during the first day, which is sustained during the next 2 days. Increased urinary excretion of potassium begins to become significant only toward the end of the first week of bed rest and continues thereafter. These net losses of fluid and sodium are believed to underlie the decreases in plasma and blood volume during the early phases of bed rest. Creatinine clearance is decreased only during the first day of bed rest. In spite of these changes in electrolyte excretion, plasma concentrations of sodium and potassium remain unchanged even with continued bed rest. However, PRA has consistently been found to increase (Chavarri et al., 1977; Dallman et al., 1984; Hargens et al., 1983; Melada et al., 1975; Volicer et al., 1976), probably in response to the sodium loss.

During bed rest, there is a marked change from normal in the coupling between PRA and aldosterone concentration, in the absence of significant changes in the levels of ACTH or electrolytes, factors that are known to alter this relationship. In three 7-day bed rest studies with male and female subjects, we have found that PRA has always been increased, whereas blood pressure remained unchanged and plasma aldosterone levels were either decreased or unchanged. Measurements of angiotensin II (A-II) did not substantiate the possible explanation that the apparent dissociation between PRA and aldosterone could have involved inhibition of angiotensin-converting enzymes (situated primarily in the lungs) resulting from the hemodynamic changes associated with the horizontal or head-down body positions. However, both PRA and aldosterone respond to changes in volume, since both show the expected reduction during a volume load (2 liters of 0.15 M NaCl infused over 4 hr) (Dallman *et al.*, 1984) or blood autotransfusion (Greenleaf, 1984). The sensitivity of adrenal aldosterone secretion in response to ACTH also appeared to be unchanged during 6 days of bed rest (Dallman *et al.*, 1984), as was renal sensitivity in response to the endogenously secreted aldosterone. Although a reduction in plasma aldosterone concentration during bed rest may have been expected to increase sodium loss and potassium retention, Chobanian and associates (1974) observed no such changes during a 21-day study. It is worth remembering here that aldosterone excretion is three times lower at night during sleep in the supine position than it is during the day when a person is upright and active. During bed rest, the day and night differences are almost absent, with the nighttime secretion sometimes being even greater (Melada *et al.*, 1975).

Plasma aldosterone levels not only vary with posture (Wolff and Torbica, 1963), but also increase rapidly after venesection, indicating that the changes involve blood volume receptors. When blood volume increases, aldosterone secretion is probably depressed by a low-pressure (right side) volume receptor–pituitary reflex and by the stretch receptor–renin involvement of the juxtaglomerular system.

The physiological fluctuations of aldosterone secretion seem to have little effect on sodium and potassium excretion. Despite the decrease in aldosterone secretion during sleep, there is no increased loss in urinary sodium, and in fact, it is reduced. In normal humans, no correlation exists between the aldosterone rhythm and the excretion of sodium. It has been suggested that although blood volume can influence aldosterone production, it also significantly and directly affects that portion of the renal system that controls sodium reabsorption and excretion (Epstein *et al.*, 1951). Furthermore, the central mechanisms regulating sodium excretion and a variety of putative natriurietic and diuretic factors have recently received a great deal of investigative attention. Their possible role in explaining some of these regulatory discrepancies during bed rest is at present unknown.

C. Catecholamine Response

Catecholamines serve not only as prime indicators of the adrenal medulla gland activity but also of the sympathetic nervous system activity level. Four-fifths of the urinary epinephrine originates in the adrenal medulla, while only 5–10% of the excreted norepinephrine has a similar origin; the rest originates in the central and autonomic nervous system (Wallin *et al.*, 1981).

The vascular system usually reponds to a decrease in circulating blood volume by releasing catecholamines. An increase in catecholamine release therefore, would be expected during bed rest, but in fact, just the opposite occurs. Moreover, no change in turnover rate during 3 weeks of horizontal bed rest was found (Chobanian *et al.*, 1974). In fact, Leach and co-workers (1973), in a study of 8 healthy volunteers bed rested for 24–30 weeks, found a significant decrease in urinary norepinephrine excretion, with no change in epinephrine. In a 182-day head-down study using daily bicycle exercise, nonexercisers showed fluctuating urinary norepinephrine levels, while as expected the exercising group excreted consistently higher norepinephrine levels during bed rest than before (Krupina *et al.*, 1980).

Several studies have now documented a significant increase in plasma norepinephrine levels with upright tilt (Fluck and Salter, 1973; Morganti *et al.*, 1979; Rosenthal *et al.*, 1978). The lack of such a release could underlie the decreased orthostatic tolerance regularly seen after bed rest. Such a possibility was supported by the work of Schmid and co-workers (1971), who found that the vasoconstrictor action of tyramine, which acts to release norepinephrine from nerve endings, was attenuated with bed rest, and from our studies showing a decreased norepinephrine release with standing after 7 days of bed rest in subjects prone to orthostatic hypotension (J. Vernikos, unpublished results). On the other hand, there had been no evidence of change in α-adrenergic responsiveness since blood pressure responses to gradient infusions of norepinephrine and A-II have remained unchanged during bed rest (Chobanian *et al.*, 1974; Schmid *et al.*, 1971).

Evidence of an increase in β-adrenergic activity with bed rest has been deduced from data on increased resting diastolic blood pressure and the increase in plasma renin activity (Melada *et al.*, 1975). Furthermore, Sandler and co-workers (1985) have recently reported that β-adrenergic blockade with propranolol provided some, but limited, benefit against orthostatic loss following bed rest.

D. Glucose Metabolism

Disordered glucose tolerance is a hallmark of inactivity, and the degree of abnormality is proportional to the degree of immobilization (Lutwak and Whedon, 1959). Hyperglycemia and hyperinsulemia in response to a glucose load have been noted during bed rest in individuals confined to wheelchairs and with varying degrees of inactivity. Research on carbohydrate metabolism has established that normal human subjects show abnormal glucose tolerance responses after prolonged inactivity (Blotner, 1945; Dolkas and Sandler, 1974; Lipman *et al.*, 1970b; Naughton and Wulff, 1967; Piemme, 1971). The condition has also been observed in patients bed rested for prolonged periods (Dolkas and Sandler, 1974; Lebovitz *et al.*, 1969). In bed rest studies with strict dietary controls, investigators have found that plasma insulin levels following a glucose load (glucose tolerance test) may increase twice as much as under normal conditions but without correspondingly dramatic changes in plasma glucose (Dolkas and Greenleaf, 1977; Lebovitz et al., 1969; Lecocq, 1971; Piemme, 1971). The hyperinsulemic response to a glucose load is apparent after only 2 days of bed rest and is more enhanced with increasing duration (Wegmann et al., 1984). Simultaneous measurements of the C-peptide that is part of the pro-insulin molecule and is released with insulin from the pancreas in equimolar amounts, have provided evidence that the excessive hyperinsulemia is caused by higher secretion rates from the pancreas and not by a diminished clearance from the circulation (Wegmann et al., 1984; Wirth et al., 1981). It would appear that insulin becomes relatively ineffective in lowering blood glucose. Similar patterns occur in the early stages of insulin-resistant diabetes and probably also occur with such clinical conditions as obesity, hyperadrenocorticism, and uremia.

In a study of subjects bed rested for 56 days, we found that mean daily resting levels of insulin increased during the first 30 days, although glucose concentrations remained unchanged, as shown in Fig. 6 (Vernikos-Danellis et al., 1974). Thereafter, insulin levels decreased toward baseline levels and glucose decreased to below normal, reaching hypoglycemic levels especially after meals. Glucose returned to normal levels during the ambulatory control period. These results suggested that under conditions of inactivity an insulin resistance develops, requiring greater amounts of insulin to maintain normal glucose levels. Since the high insulin secretion is not or cannot be maintained, glucose levels decrease and could reach hypoglycemic concentrations.

Glucose intolerance is frequently observed in patients with spinal cord injuries. Duckworth and associates (1983) found that 23% of the patients discharged from a spinal cord service experienced fasting hyperglycemia. In their screening

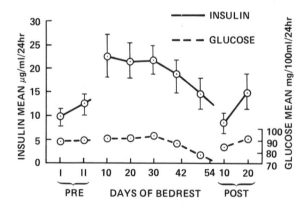

Fig. 6. Mean circulating insulin (solid line) and glucose (broken line) levels per 48-hr sampling period before, during, and after 56 days of bed rest; vertical lines represent standard errors ($N = 5$).

study, they found that 30% of a group of stable spinal cord injury patients had 2 hr glucose values exceeding 200 mg/dl following a glucose load and that 40% had values exceeding 140 mg/dl. In many such patients, glucose intolerance appears as hyperinsulemia after glucose is administered either orally or intravenously. In addition, more than 40% of a group of glucose-intolerant patients also had elevated insulin levels, suggesting that a tissue resistance to endogenous insulin occurs.

Although it is not known precisely why insulin resistance occurs with bed rest, it could result from either a primary defect in insulin-sensitive tissues or the presence of circulating insulin antagonists. Some investigators have noted, however, that the usual antagonists—growth hormone, catecholamines, cortisol, and free fatty acids—probably are not responsible because they do not increase significantly or may even be decreased during prolonged bed rest (Cardus et al., 1965; Lipman et al., 1970a,b). It has also been suggested that the sera of bed rested subjects may contain as yet unidentified circulating insulin antagonists that alter the response of the liver to elevated insulin levels (Dolkas and Greenleaf, 1977). Alternatively, insulin resistance may result from a defect in insulin binding and insulin receptor sensitivity.

There is also some thought that the metabolic changes in muscle that occur with prolonged physical inactivity may contribute to insulin resistance and changes in glucose uptake. Lipman and associates (1970a,b) found a diminished peripheral uptake of glucose by the muscles of the forearm and proposed that altered muscle uptake was partially responsible for the decrease in glucose tolerance observed during bed rest. Ellis and associates (1974) noted that plasma alanine concentration increased significantly in bed rested subjects and suggested that changes in the amino acid metabolism of the larger, more gravity-dependent muscles of the legs and back probably were responsible for the higher level of alanine in arterial blood and that the higher venous level observed was a passive consequence of the higher arterial level. Portugalov and associates (1971) reported that the enzyme content of skeletal muscle of immobilized rats changed as early as the first day of immobilization.

The importance of activity in maintaining a normal insulin–glucose relationship has also been emphasized by work with athletes, using exercise to reverse some of the effects of bed rest. Daily isometric or isotonic exercise decreased the hyerinsulemic response to a glucose load but did not eliminate it. Isotonic exercise was more effective than isometric in this regard (Dolkas and Greenleaf, 1977). When subjects were again ambulatory, 7–14 days were required to restore the plasma glucose level to normal in nonexercising individuals, and 7 days in those who exercised during bed rest (Greenleaf and Kozlowski, 1982). More recently, Wegmann and associates (1984) have compared responses to a glucose load in highly trained athletes ($\dot{V}_{O_2\ max}$ of 70 ml/kg body weight) and untrained, healthy volunteers ($\dot{V}_{O_2\ max}$ of 40 ml/kg body weight) during 7 days of $-6°$

head-down bed rest. The athletes showed distinctly lower insulin responses to the standard glucose load than did their untrained counterparts. Moreover, the kinetics of both the glucose and insulin responses were different in the two groups, suggesting that higher physical fitness results in a delayed and slower adjustment to glucose loading and a greater tendency to reach hyperglycemia at the end of the test.

In both animals and humans, physical training and inactivity are obviously major factors in the regulation of glucose by insulin and should be taken into consideration when the glucose tolerance test is used for diagnostic purposes. Similarly, although the mechanism for the development of insulin resistance during bed rest is not understood, the insulin–glucose changes that occur with inactivity should be of special concern in clinical treatment involving immobilization. The increased circulating levels of insulin required to maintain normal glucose levels; the fluctuations in growth hormone, insulin, and glucose; and the occurrence of periodic hypoglycemia (possibly in response to meals) all indicate that prolonged bed rest results in an impairment of the mechanism regulating glucose homeostasis and creates serious questions about the advisability of prolonged bed rest and the hazards of an inactive life.

II. POST-BED-REST ORTHOSTATIC HYPOTENSION AND COUNTERMEASURES

Those who get out of bed after a period of inactivity or bed rest and those who return to earth's gravity (1g) after spaceflight commonly experience orthostatic hypotension, with some being more susceptible than others. Normally, when an individual stands up, there is a slight reduction in systolic arterial pressure, which triggers a series of physiological events designed to adjust and to maintain adequate perfusion to the brain. Stimulation of baroreceptors causes autonomic nervous system activity that results in increased peripheral constriction of both arteries and veins; the heart rate and the contractility of the heart muscle are also increased. Norepinephrine and AVP are released, and activation of the renin-angiotensin-aldosterone and pituitary adrenocortical mechanisms help maintain intravascular volume and pressure. If all or some of those responses do not occur or are inadequate, the pull of gravity in the head-to-toe axis would cause excessive pooling of blood in the peripheral veins, resulting in a substantial reduction in central blood volume, left ventricular filling pressure, cardiac output, and finally cerebral blood flow, and fainting would result.

Moreover, vasopressor mechanisms may be involved in which sudden, excessive autonomic (vagal cholinergic) activity causes the arteries and arterioles to dilate, with bradycardia, sweating, nausea, and hypotension following. Reduced effective circulating blood volume during bed rest resulting from shifts in fluids and electrolytes and reduced vasoreactivity can also contribute to orthostatic hypotension following spaceflight or bed rest.

Changes in fluid and electrolyte balance have been seen consistently in both bed rest and spaceflight, with orthostatic hypotension occurring upon return to ambulation or to earth's gravity. After bed rest, hormone and neurohumoral responses to orthostatic stimuli are enhanced, but individuals who are susceptible to orthostatic hypotension show inadequate increases in circulating norepinephrine; these responses are sometimes seen in susceptible individuals even prior to bed rest. Responses have been similar in both male and female subjects. We have hypothesized that the problem results primarily from a decreased ability to release norepinephrine from sympathetic nerve endings, which is aggravated by bed rest (J. Vernikos, in preparation). Much research has gone into developing countermeasures to the changes observed with bed rest and spaceflight. The use of isometric and isotonic exercise during bed rest has been studied frequently (Greenleaf et al., 1982) as have lower body negative pressure (Sandler, 1980), electrostimulation of muscles (Cherepakhin et al., 1977), hypoxia, local cooling of the legs (Keatinge and Howard, 1971; Raven et al., 1980), and pressure applied to the lower half of the body to reduce venous pooling (Sheps, 1976). Pressure application has been the most effective of these physical measures in partially reducing orthostatic intolerance after spaceflight or bed rest and even in the management of clinical problems resulting from immobilization.

Pharmacologic agents have also been investigated as countermeasures. Ingestion of fluids has been used to improve plasma volume in both spaceflight (Pool and Nicogossian, 1983) and bed rest (Greenleaf et al., 1973; Hyatt and West, 1977). But this measure has had limited beneficial effects. Clinically, the most effective single drug has been fludrocortisone (9-α-fluorohydrocortisone) (Schatz et al., 1976). Why this drug works to offset orthostatic hypotension is not understood. It does expand blood volume at first, but the effect is transient and in some individuals it is minor. The observed increase in blood pressure cannot be attributed entirely to increased plasma volume. It has been suggested that other mechanisms may be at work, including the drug's ability to intensify the vascular actions of circulating norepinephrine, to release a vasopressor principle, or to change vascular electrolyte concentrations, thus enhancing vasoreactivity. Bohn and associates (1970) reported that daily administration of this drug increased plasma volume and improved orthostatic tolerance after 10 days of horizontal bed rest. This conclusion, however, differed from the results of Stevens and associates (1966), who found in a 30-day bed rest study that the use of the drug combined with the inflation of venous-occlusive cuffs for 6 hr per day on the last 2–4 days of bed rest had little effect on orthostatic hypotension even though the procedure restored plasma volume to normal. Hyatt (1971) found similar results with 70° head-up tilt. It may be that different mechanisms come into play with more prolonged periods of bed rest, or perhaps the drug dosages or duration of treatment were insufficient. Massive doses of cortisol are recommended for preoperative preparation and for Addisonian crises (Felig and Tamborlane, 1981). Cortisol has restored plasma volume and proteins in dogs following hemorrhage (Pirckle and Gann, 1975). Both cortisol and 9-α-fluorohydrocortisone have equilavent affinity for the Type I glucocorticoid receptor. Although the prolonged use of corticosteroids may be inadvisable, the information to date on their effect on post-bed-rest orthostatic hypotension would suggest that it should be investigated further.

The part played by β-adrenergic function in orthostatic hypotension is not yet clear. Melada and associates (1975) reported that IV propranolol had beneficial effects after bed rest, and this finding was confirmed recently by Sandler and coworkers (1985). After bed rest, systemic vascular resistance was maintained during tilt with propranolol, despite the expected fall in cardiac output, indicating either a blockade of β_2-adrenergic vasodilation or a release of epinephrine from the adrenals. Such possibilities need further study. Administration of atropine, to date, has shown no overwhelmingly positive protective effects. Similarly, Soviet investigators have had only limited success with the calcium blocking agent, isoptin (Bogolyubov et al., 1978). Some beneficial effects have been reported, however, from the administration of the central α-noradrenergic stimulant clonidine following bed rest (Guell et al., 1982).

The ideal drug for use as a countermeasure should mimic the effects of norepinephrine as it is released from sympathetic postglangionic nerves, causing selective vasoconstriction when the body is upright, in order to augment central blood volume and cardiac output. Since the maintenance of blood pressure in the upright position depends more on an adequate volume than on the loss of baroreceptor control, an expansion of effective blood volume would also be desirable (Bannister et al., 1979). In most cases, post-bed-rest orthostatic hypotension is transient, and no drastic, long-term measures would be necessary. Short-term benefits have been reported using phenylephrine, ephedrine, and metaraminol. Noradrenergic activity could also be enhanced by drugs that stimulate the release of norepinephrine from sympathetic nerve endings; tyramine and amphetamine are best known for such purposes. Tyramine has been administered clinically either in pure form or in tyramine-containing foods (e.g., cheddar cheese). In the case of tyramine, the action is intensified by the administration of a monamine oxidase inhibitor that interferes with norepinephrine metabolism (Diamond and Buchanan, 1970). Clinical trials of this combination, however, have resulted in conflicting findings (Bannister et al., 1979; Nanda Kumar et al., 1976).

More recently, dietary manipulations have been reported to enhance the activity of sympathetic neuronal discharge. They include a low-sodium diet (Gordon et al., 1967; Robertson et al., 1979), increased caffeine intake (Robertson et al., 1978), and a diet high in carbohydrates. Although severe restriction of sodium to 10 mEq of dietary sodium per day selectively increases both supine and ambulatory levels of norepinephrine, the increase is relatively modest. Moreover, such a diet could not be used where there is an existing natriuresis. Although ingestion of caffeine (250 mg) does increase norepinephrine levels, it acts primarily on the adrenal medulla so that epinephrine levels are increased by more than twofold (Robertson *et al.*, 1978). Hence, caffeine would not be expected to be an effective countermeasure because orthostasis normally does not involve significant stimulation of the adreno- medullary function. Reports on the effect of carbohydrate-induced changes in sympathetic nervous activity in humans and animals (Landsberg and Young, 1981; Welle *et al.*, 1981) may have more practical implications. In a study using both young and elderly men, Young and associates (1980) found that the administration of oral glucose increased plasma norepinephrine significantly. The effect seems to be specific to carbohydrate intake, since Welle and associates (1981) found that neither high protein nor high fat consumption had such an effect.

As mentioned previously, atropine has also been considered as a countermeasure, but results have been conflicting (Bjurstedt *et al.*, 1977; Stegemann *et al.*, 1969). However, combined with other procedures designed to enhance noradrenergic activity, it may be particularly useful.

III. DRUG METABOLISM AND DRUG ACTION

An altered metabolic state will invariably affect the rate at which drugs are absorbed, distributed, bound, metabolized, or excreted. This, in turn, will affect the availability and effectiveness of a medication. Although the possibility that genetics, disease states, or interaction with other medications may alter drug metabolism has received a great deal of attention in recent years, the concept that the level of activity in healthy individuals should be considered has not received the attention it deserves. However, we know that the process of athletic conditioning results, at the very least, in increased plasma volume and that bed-rest-associated inactivity results in a restriction of this volume, which would be expected to affect circulating drug levels.

Once more, space medicine required answers that were not otherwise available. Investigators began to question how humans would react when required to take medications in weightlessness. How would the drugs be metabolized, and how would the recipients respond? Could a drug dose normally handled on earth be overwhelming in space, or would it be ineffective? Answers to such questions are important not only for space crews, but also for those exposed to other forms of inactivity, such as prolonged bed rest, hospitalization, clinical immobilization, and aging.

Even the acute changes in plasma volume brought about by standing up or lying down will alter the kinetics of theophylline. This drug is highly soluble in body fluids and typically exhibits a two-compartment distribution (Riegelman *et al.*, 1968). That is, after administration, the plasma level of the drug peaks as the drug is absorbed into the circulating volume and stops as it equilibrates with the extravascular second compartment. It has also been shown that aminophylline, when infused in normal standing subjects, results in significantly higher peak plasma levels than it does when they lie down (Warren *et al.*, 1983). In the upright posture, plasma volume can drop by 5–15% (Dallman *et al.*, 1984). Thus, aminophylline from slow-release preparations would reach lower peak plasma levels in supine patients at night than in ambulatory individuals in the daytime and may explain why fewer side effects to the drug are seen when it is administered at night in the supine position.

In 1967, Levy reported that bed rest resulted in an increase in renal clearance of penicillin but a reduction in its metabolic clearance. On the other hand, Kates and associates (1980) found no significant changes in the clearance of penicillin and lidocaine in 12 men bed rested for 14 days. Penicillin (1 million units) and lidocaine (100 mg) were administered intravenously twice during an ambulatory control period and on days 9 and 10 of bed rest. Penicillin was given as a rapid bolus, whereas lidocaine was infused over 15 min; blood samples were taken before and at frequent intervals after administration of the drugs. These drugs were selected for study because they are eliminated primarily through hepatic and renal routes. The investigators found no increase in plasma concentration of the drugs, which was unexpected. Both of these drugs have small distribution volumes, so the reduced plasma volume seen consistently with bed rest should have resulted in increased concentration of the drugs. It was concluded that the blood flow and hepatic function did not change sufficiently during bed rest to alter drug metabolism. Thus, it would appear that the physiological changes during bed rest may not affect drug distribution or elimination significantly. Soviet investigators have reached similar general conclusions through indirect measurements of the cardiovascular effects of some central nervous system stimulants (Belay and Uglova, 1974).

Whereas activity or inactivity may alter receptor number or sensitivity, the response to specific drugs or drug categories should be predictable. Sensitivity to insulin may be one such example (see Section I,C). Drugs that act on skeletal muscle offer another example. Both reduced blood flow and disuse atrophy are known to alter the sensitivity of the neuromuscular junction to muscle relaxants. Resistance to nondepolarizing muscle relaxants, such as pancuronium and D-tubocurarine, during inactivity have been reported. Gronert (1981) reported a resistance to pancuronium in the immobilized extremities of a dog, which did not occur in the nonimmobilized parts. A similar resistance to nondepolarizing muscle relaxants observed in the paretic upper extremities of patients with residual hemiplegia (Moorthy and Hilgenberg, 1980) or upper motor neuron lesions (Graham, 1980) is suggestive of altered pharmacodynamics rather than changes in pharmacokinetics. Gronert (1981) argues that disuse atrophy during immobilization results in an enlargement of the muscle end plate area and an increase in muscle membrane sensitivity to acetylcholine: "Thus, the more numerous receptor sites would bind more molecules of relaxant, necessitating a greater total dose of the drug." On the other hand, there is considerable evidence now that for similar reasons, inactivity increases the sensitivity to depolarizing muscle relaxants, such as succinylcholine. This is manifested by a hyperkalemic response to succinylcholine which has occasionally been fatal. Thus, during the treatment of the inactive or bedridden patient careful consideration must be given to the possible consequences on the pharmacokinetics or pharmacodynamics of the medication.

IV. CONCLUSIONS

Although there are a great number of reported metabolic changes in individuals who are inactive by virtue of age, disease, disability, or spaceflight, the contribution of inactivity to these overall changes cannot be reliably isolated. Hence, to increase reliability, the data discussed here have depended primarily on studies using healthy individuals in whom the level of inactivity could be controlled by bed rest. Nevertheless, a pattern evolves common to all of these conditions, which is characterized by early and rapid changes in fluids and electrolytes and a reduction in blood volume, which, with continued inactivity, is followed (a) by uncoupling in many neuroendocrine and neurohumoral regulatory mechanisms and (b) by changes in the number and affinity of receptors, such as those to insulin and the neuromuscular junction. The extent of these changes seems to vary with the original physiological make-up of the individual (e.g., athletic versus nonathletic, fainter versus nonfainter). Maintenance or reversibility of these changes during inactivity can only come from careful identification of the mechanisms involved. There is no doubt, however, that a healthy individual made inactive for even relatively short periods of time—and reversible as the responses might be—develops a very dramatic and characteristic syndrome.

REFERENCES

Bannister, R., Davies, B., Holly, E., Rosenthal, T., and Sever, P. (1979). *Brain* **102(1)**, 163-176.
Belay, V. Ye., and Uglova, N.N. (1974). *Kosmich. Biol. Aviak. Med.* **8**, 83-84.
Bistrian, B.R., Blackburn, G.L., Scrimshaw, N.S., and Flatt, J.P. (1975). *Amer. J. Clin. Nutr.* **28**, 1148.
Bjurstedt, H., Rosenhamer, G., and Tyden, G. (1977). *Acta. Physiol. Scand.* **99**, 353-360.
Blomqvist, C.G., Nixon, J.V., Johnson, R.J., Jr., and Mitchell, J.H. (1980). *Acta Astronautica* **7**, 543-553.

Blotner, H. (1945). *Arch. Intern. Med.* **75**, 39-44.

Bogolyubov, V.M., Anashkin, O.D., Trushinskiy, Z.K., Sashkov, V.S., Shatunia, T.P., and Reva, F.V. (1978). *Voyenno-Meditsinskiy Zhurnal* **11**, 64-66.

Bohn, B.J., Hyatt, K.H., Kamenetsky, L.G., Calder, B.E., and Smith, W.M. (1970). *Aerospace Med.* **41**, 495-499.

Browse, N. L. (1965). *Physiology and Pathology of Bed Rest.* Charles C. Thomas, Springfield, Il., 221 pp.

Butterworth, C.E., Jr. (1974). *J.A.M.A.* **230**, 879.

Cardus, D., Vallbona, C., Vogt, F.B., Spencer, W.A., Lipscomb, H.S. and Eik-Nes, K.B. (1965). *Aerospace Med.* **36**, 524-528.

Chavarri, M., Ganguly, A., Luetscher, J.A., and Zager, P.G. (1977). *Aviat. Space Environ. Med.* **48**, 633-636.

Cherepakhin, M. A., Kakurin, L. I., Il'ina-Kakuyeva, Ye. I., and Fedorenko, G.T. (1977). *Space Biol. and Aerospace Med.* **11(2)**, 64-68.

Chobanian, A.V., Lille, R.D., Tercyak, A., and Blevins, P. (1974). *Circulation* **49**, 551-559.

Consolazio, C.F., Johnson, H.L., Krzywicki, H.J., and Witt, N.F. (1972). *Amer. J. Clin. Nutr.* **25**, 572

Dallman, M.F., Keil, L.C., Convertino, V., O'Hara, D., and Vernikos-Danellis, J. (1984). In: *Stress: The Role of Catecholamines and Other Neurotransmitters* (E. Usdin, R. Kvetnansky, and J. Axelrod, eds.) Vol. II, pp. 1057-1077. Gordon and Breach Science Publishers, London.

Deitrick, J.E., Whedon, G.D., and Shorr, E. (1948). *Amer. J. Med.* **4**, 3-35.

Diamond, W.D., and Buchanan, R.A. (1970). *J. Clin. Pharmacol.* **10**, 306-311.

Dolkas, C.B., and Sandler, H. (1974). Countermeasure effectiveness of abnormal glucose tolerance during bedrest. *Aerospace Med. Assoc. Preprints*, pp. 169-170.

Dolkas, C.B. and Greenleaf, J.E. (1977). *J. Appl. Physiol.: Respirat. Environ. Exercise Physiol.* **43(6)**, 1033-1038.

Donaldson, C.L., Hulley, S.B., McMillian, D.E., Hattner, R.S., and Bayers, J.H. (1969). "The Effect of Prolonged Simulated Non-gravitational Environment on Mineral Balance in the Adult Male." NASA Contractor Report 108314, pp. 1-91.

Duckworth, W.C., Jallepalli, P., and Solomon, S.S. (1983). *Arch. Phys. Med. Rehabil.* **64(3)**, 107-110.

Ellis, J.P., Lecocq, F.R., Garcia, J.B., and Lipman, R.L. (1974). *Aerospace Med.* **45**, 15-18.

Epstein, H.F., Goodyer, A.V.N., Lawrason, F.D., and Relman, A.S. (1951). *J. Clin. Invest.* **30**, 63.

Felig, P., and Tamborlane, W.V. (1981). *Ann. Intern. Med.* **93(4)**, 627-629.

Fluck, D.C. and Salter, C. (1973). *Cardiovasc. Res.* **7**, 823-826.

Gordon, M.W., Deanin, G.G., and Hanson, R.K. (1967). *Biochem. Pharmacol.* **16**, 793-802.

Graham, D.H. (1980). *Anesthesiology* **52(1)**, 74-75.

Greenleaf, J.E. (1984). *J. Appl. Physiol: Resp. Em.-Exc.* **57**, 619-633.

Greenleaf, J.E. and Kozlowski, S. (1982). *In* "Exercise and Sport Sciences Reviews" (R.A. Terjung, ed.). Vol. 10, pp. 84-119. Franklin Institute Press, Philadelphia.

Greenleaf, J.E., Van Beaumont, W., Bernauer, E.M., Haines, R.F., Sandler, H., Staley, R.W., Young, H.L., and Yusken, J.W. (1973). *Aerospace Med.* **44**, 715-722

Greenleaf, J.E., Bernauer, E.M., Young, H.L., Morse, J.T., Staley, R.W., Juhos, L.T., and Van Beaumont, W. (1977). *J. Appl. Physiol.* **42**, 59-66.

Greenleaf, J.E., Silverstein, L., Bliss, J., Langenheim, V., Rossow, H., and Chao, C. (1982). " Physiological Responses to Prolonged Bedrest and Fluid Immersion in Man: A Compendium of Research (1974-1980)." NASA Tech Memo 81324, pp. 1-110.

Gronert, G.A. (1981). *Anesthesiology* **52(4)**, 356.

Guell, A., Gharib, Cl., Fanjaud, G., Dupui Ph. and Bes, A. (1982). *Physiologist* **25(4), S71-S72.**

Hargens, A.G., Tipton, C.M., Gollnick, P.D., Mubarak, S.J., Tucker, B.J., and Akeson, W.H. (1983). *J. Appl. Physiol.* **54**, 1003-1009.

Hodges, R.E. (1980). "Nutrition in Medical Practice." W.B. Saunders Company, Philadelphia. 366 pp.

Hyatt, K.H. (1971). In "Hypogravic and Hypodynamic Environments" (R.H. Murray and M. McCally, eds.). NASA SP-269, pp. 187-209.

Hyatt, K.H., and West, D.A. (1977). *Aviat. Space Environ. Med.* **48**, 120-124.

Jacobson, L.B., Hyatt, K.H., Sullivan, R.W., Sandler, H., Rositano, S.A., and Mancini, R. (1973). "Evaluation of $+G_z$ Tolerance Following Simulated Weightlessness (Bedrest)." NASA TM X-62311, 75 pp.

Kaplan, L., Powell, B.R., Grunbaum, B.B., and Rusk, H.A. (1966). "Comprehensive Follow-Up Study of Spinal Cord Dysfunction and Its Resultant Disabilities." Institute Medical Rehabilitation, New York, pp. 184.

Kates, R.E., Harapat, S.R., Keefe, D.L.D., Goldwater, D., and Harrison, D.C. (1980) *Clin. Pharmacol. Ther.* **28(5)**, 624-628.

Katz, F.H., Romfh, P., and Smith, J.A. (1975). *J. Clin. Endocrinol. Metab.* **40**, 1049-1055.

Keatinge, W.R., and Howard, P. (1971). *J. Appl. Physiol.* **31**, 819-822.

Kimzey, S.L., Leonard, J.T., and Johnson, P.C. (1979). *Acta Astronautica* **6**, 1289-1303.

Krotov, V.P., Titov, A.A., Kovalenko, E.A., Bogomolov, V.V., Stazhadze, L.L., and Masenko, V.P. (1977). *Kosmich. Aviakosmich. Med.* **1(32)**.

Krupina, T.N., Iarullin, Kh.Kh., and Alekseev, D.A. (1980). *Kosm. Biol. Aviakosm. Med.* **14(2)**, 49-54.

Landsburg, L., and Young, J.B. (1981). *Int. J. Obes.* **1**, 79-91.

Law, D.K., Dudrick, S.J., and Abdou, N.I. (1974). *Surg. Gynecol. Obstet.* **139**, 257.

Leach, C.S., Hulley, S.B., Rambaut, P.C., and Dietlein, L.F. (1973). *Space Life Sci.* **4**, 415-422.

Leach C.S., Vernikos-Danellis, J., Winget, C.M., Rambaut, P.C., and Campbell, B.O. (1974). In "Biorhythms and Human Reproduction" (M. Ferin, F. Halberg, R.H. Richart, and R.L. Vande Wiele, eds.). Wiley, New York, pp. 406-416.

Lebovitz, H.E., Shultz, K.T., Matthews, M.E., *et al.* (1969). *Circulation* **39**, 171-181.

Lecocq, F.R. (1971). "The Effect of Bedrest on Glucose Regulation in Man." NASA SP-269, p. 288.

Levy, G. (1967). *J. Pharmacol. Sci.* **56**, 928-929.

Lipman, R.L., Schnure, J.J., Bradley, E.M., and Lecocq, F.R. (1970a). *J. Lab. Clin. Med.* **76**, 221-230.

Lipman, R.L., Ulvedal, F., Schnure, J.J., Bradley, E.M., and Lecocq, F.R. (1970b). *Metabolism* **19**, 980-987.

Lutwak, L., and Whedon, G.D. (1959). *Clin. Res.* **7**, 143-144.

Melada, G.A., Goldman, R.H., Luetscher, J.A., and Zager, P.G. (1975). *Aviat. Space Environ. Med.* **46**, 1049-1055.

Miller, P.B., Johnson, R.L., and Lamb, L.E. (1965). *Aerospace Med.* **36**, 1077-1082.

Moorthy, S.S. and Hilgenberg, J.C. (1980). *Anesth. Analg.* (Cleve.) **59(8)**, 624-627.

Morganti, A., Lopez-Ovejero, J.A., Pickering, T.G., and Laraph, J.H. (1979). *Am. J. Cardiol.* **43**, 600-604.

Munster, A.M. (1976). *Lancet* **1**, 1329.

Nanda Kumar, N.V., Visweswariah, K., and Majumber, S.K. (1976). *J. Assoc. Off. Anal. Chem.* **59(3)**, 641-643.

Naughton, J., and Wulff, J. (1967). *J. Lab. Med.* **70**, 996.

Pace, N., Kodama, A.M., Price, D.C., Grunbaum, B.W., Rahlmann, D.F., and Newsom, B.D. (1976). *Life Sci. Space Res.* **14**, 269-274.

Piemme, T.E. (1971). *In* "Hypogravic and Hypodynamic Environments" (R.H. Murray and M. McCally, eds.) NASA SP 269, pp. 281-287.

120

Joan Vernikos

Pirkle, J.C., Jr. and Gann, D.S. (1975). *Am. J. Physiol.* **228(3)**, 821-827.

Pool, S.L., and Nicogossian, A.E. (1983). *Aviat. Space and Environ. Med.* **54** *(Suppl. l)*, 541-549.

Portugalov, V.V., Ilyina-Kaueva, E.I., Starostin, V.I., Rokhlenko, K.D., and Savick, Z. F. (1971). *Aerospace Med.* **42**, 1041-1049.

Raven, P.B., Saito, M., Gaffney, F.A., Schutte, J., and Blomqvist, C.G. (1980). *Aviat. Space Environ. Med.* **51**, 497-503.

Riegelman, S., Loo, J.C., and Rowland, M. (1968). *J. Pharm. Sci.* **57**, 117-123.

Robertson, D., Frohlich, J.C., Carr, R.K., Watson, J.T., Hollifield, J.W., Shand, D.G., and Oates, J.A. (1978). *N. Eng. J. Med.* **298(4)**, 181-186.

Robertson, D., Robertson, R.M., Nies, A.S., Oates, J.A., and Friesinger, G.C. (1979). *Am. J. Cardiol.* **43(6)**, 1080-1085.

Rosenthal, T., Birch, M., Osikowska, B. and Sever, P.S. (1978). *Cardiovasc. Res.* **12**, 144-147.

Sandler, H. (1980). *In* "Hearts and Heart-Like Organs" (G.H. Bourne, ed.), Vol. 2, pp. 435-524. Academic Press, New York.

Sandler, H. and Winter, D.L. (1978). "Physiological Responses of Women to Simulated Weightlessness. A Review of the Significant Findings of the First Female Bed Rest Study." NASA SP-430, pp. 1-87.

Sandler, H., Goldwater, D.J., Popp, R.L., Scapavento, L., and Harrison, D.C. (1985). *Am. J. Cardiol.* **55**.

Schatz, H., Nierle, C., and Pfeiffer, E.F. (1976). *Eur. J. Clin. Invest.* **5(6)**, 477-485.

Schmid, P.G., Shaver, J.A., McCally, M., Bensy, J.J., Pawlson, L.G., and Pienume, T.E. (1971). *Aerosp. Med. Assoc. Preprints*, p. 104.

Sheps, D.S. (1976). *In* "Self Assessment in Clinical Cardiology" (M.S. Gordon, ed.) Vol. II. Year Book Medical Publishers, Chicago. pp. 104-110.

Stegemann, J., Framing, H.D., and Schiefeling, M. (1969). *Pflugers Arch.* **312**, 129-138.

Steinberg, F.U. (1980). "The Immobilized Patient: Functional Pathology and Management." Plenum Medical Book Company, New York. 156 pp.

Stevens, P.M., Miller, P.B., Lynch, T.N., Gilbert, C.A., Johnson, R.L., and Lamb, L.E. (1966). *Aerospace Med.* **37**, 466-474.

Taylor, H.L., Erickson, L., Henschel, A., and Keys, A. (1945). *Amer. J. Physiol.* **144**, 227-232.

Vallbona, C., Lipscomb, H.S., and Carter, R.E. (1966). *Arch. Phys. Med. Rehab.* **47**, 412-421.

Vendt, V. P., Kindrat'yeva, L.G., Govseyeva, N.N., Apukhovskaya, L.I., Ivashkevich, S.P., Koval, V.G., and Tigranyan, R.A. (1979). *Kosmich. Biol. Aviak. Med.* **13**, 43-47.

Vernikos, J. (1984). In "Breakdown in Human Adaptation to Stress. Vol. l, Part 2. Human Performance and Breakdown in Adaptation" (H.M. Wegmann, ed.), Kluwer Academic Publishers, Boston, pp. 509-521.

Vernikos-Danellis, J., Winget, C.M., Leach, C.S., and Rambaut, P.C. (1974). "Circadian, Endocrine, and Metabolic Effects of Prolonged Bedrest: Two 56-Day Bedrest Studies." NASA TM X-3051.

Vernikos-Danellis, J., Dallman, M., Forsham, A.L., Goodwin, A.L., and Leach, C.S. (1978). *Aviat. Space and Environ. Med.* **49**, 886-889.

Volicer, L., Jean-Charles, R., and Chobanian, A.V. (1976). *Aviat. Space Environ. Med.* **47**, 1065-1068.

Wallin, B.G., Sundlof, G., Eriksson, B.M., Dominiak, P., Grobecker, H., and Lindblad, L.E. (1981). *Acta Physiol. Scand.* **111**, 69-73.

Warren, J.B., Dalton, N., and Turner, C. (1983). *Br. J. Clin. Pharmacol.* **16**, 405-411.

Wegmann, H.M., Baisch, F., and Schafer, G. E. (1984). Proc. Aerospace Medical Meeting, San Diego, CA.

Welle, S., Lilavivathana, U., and Campbell, R.G. (1981). *Metabolism* **29(9)**, 806-809.

White, P.D., Nyberg, J.W., Finney, L.M., and White, W.J. (1966). "Influence of Periodic Centrifugation on Cardiovascular Functions of Men During Bed Rest." NASA CR-65422, Douglas Aircraft Co., Inc., Santa Monica, CA, pp. 1-54.

Wirth, A., Diehm, C., Mayer, H., Morl, H. Vogel, I., Bjorntorp, P., and Schlief, G. (1981). *J. App. Physiol.* **50**, 71-77.

Wolff, H.P., and Torbica, M. (1963). *Lancet* **1**, 1346.

Young, H.L., Juhos, L., Castle, L., Yusken, J., and Greenleaf, J.E. (1973). "Body Water Compartments During Bed Rest." NASA Tech Report R-406.

Young, J.B., Rowe, J.W., Pallotte, J.A., Sperron, D., and Landsberg, L. (1980). *Metabolism* **29**, 532-539.

6

Psychosocial and Chronophysiological Effects Of Inactivity and Immobilization

CHARLES M. WINGET AND
CHARLES W. DEROSHIA
Biomedical Research Division
National Aeronautics and Space Administration
Ames Research Center
Moffett Field, California 94035

I. INTRODUCTION

Several environmental factors associated with the space operational environment can be simulated in ground-based immobilization studies. These factors include confinement and isolation from social contact, the exclusion or reduction of environmental synchronizer influences, and hypogravity. This chapter focuses on the results of studies to determine the effects of these factors on human psychosocial and chronophysiological adaptability.

Confinement is characterized by restraints placed upon an individual or a group so as to limit mobility or action or both (Smith, 1969). Isolation is obviously relative and involves either sensory, perceptual, or social deprivation (Fraser, 1966) and refers to separation from a "normal" environment. Several investigators (Chambers and Fried, 1963; Ormiston, 1961; Sells and Berry, 1961; Walters and Henning, 1961; Wheaton, 1959) suggest that isolation refers to the separation of an individual from his or her normal environment and may occur concomitant with confinement. Perceptual isolation involves disturbances of perception, as demonstrated by the McGill University experiments (Bexton *et al.*, 1954), and sensory distortion, as noted by Zubek (1973). Social isolation incorporates both solitude and separation from a normal social environment and may result in the attenuation of social stimuli. Rarely do confinement and isolation exist as single entities.

123

Thus, the space operational environment is a socially attenuated, physically determinative (allowing free movement within a confined area, as opposed to physically restrictive, which uses physical restraints to confine an individual), and temporally restrictive/determinative environment (because of work schedules and mission duration, respectively; Fraser, 1966). This environment does not represent strict isolation, since social interaction persists between crew members and there is outside radio communication. In this chapter, the terms "isolation" and "confinement" will denote social isolation and physically determinative and temporally restrictive/determinative confinement.

Another factor associated with the space mission environment is the reduction in the effectiveness of periodic environmental synchronizers that are necessary for maintaining sleep/wake and other circadian rhythms. This reduction in synchronizer strength may result from (a) confinement and reduced social contact, (b) elimination of the cyclic alteration between the orientation of the transverse and longitudinal axes of the body to gravitational force (zero gravity), (c) reduced light intensity, and (d) increase in work load stressors. Many studies of the effects of isolation have used bed rest or immersion to simulate the long-term effects of weightlessness. Furthermore, their experimental designs have incorporated one or more of these variables responsible for environmental synchronization. Several isolation studies have concentrated only on the effects of isolation upon performance and behavior (Fraser, 1966; Smith, 1969). Other studies have combined isolation with the exclusion of environmental synchronizers [e.g., constant light (Czeisler, 1978; Wever, 1979; Winget et al., 1976)] or altered rest/activity schedules (Wever, 1979; Winget et al., 1976). Although water immersion, chair restraint, immobilization, and bed rest have all been used to simulate weightlessness, none provide an exact replication, since gravity obviously is still present. However, this approach results in physiological changes very similar to those observed during or following weightlessness exposure. These changes include muscle atrophy, bone demineralization, redistribution of body fluids, and cardiovascular deconditioning (Calvin and Gazenko, 1975; Murray and McCally, 1971; Nicogossian and Parker, 1982).

Bed rest may be horizontal or antiorthostatic (head-down tilt). Headdown bed rest reproduces the early physiological effects of weightlessness more closely than horizontal bed rest (Kakurin et al., 1976). It is evident that all human bed rest studies subject individuals to some degree of confinement and isolation from normal social contacts in addition to the simulation of the physiological effects of weightlessness. In some bed rest experiments, the subjects may also experience sensory deprivation (Campbell, 1984).

The ability to adapt to bed rest isolation conditions, as well as to immobilization or inactivity, can be affected by age (Helmreich, 1966); respect for rules and regulations (Gunderson et al., 1964); a low need for avocational or physical activity (Myers et al., 1966); the tendency to psychosocial complaints (Eilbert and Glaser, 1959); marital status (Epstein, 1962); need for change (Hull and Zubek, 1962); emotional maturity (Weybrew, 1957); energy level, impulse control, the quality of interpersonal relationships, and aggressiveness (Weybrew, 1957); Minnesota Multiphasic Personality Inventory (MMPI) scores (Myers *et al.*, 1966); and the degree of psychopathology and existence of interpersonal problems (Eilbert and Glaser, 1959; Weybrew, 1971).

II. ISOLATION

People's adaptability to confinement or isolation has been examined under several conditions: (a) antarctic/arctic stations (Edholm and Gunderson, 1973; Gunderson, 1974); (b) fallout shelters (Baker and Rohrer, 1960; Hammes, 1965; Honigfeld, 1965); (c) laboratory studies (Rasmussen, 1973; Zubek *et al.*, 1969a,b); (d) spaceflight simulation studies (Chambers, 1964; Chambers and Fried, 1963); (e) submarine duty (Weybrew, 1971); and (f) underwater habitats (Helmreich, 1971; May, 1970; Shulte, 1967). Fraser (1966) and Smith (1969) found that confined and isolated subjects became irritable and hostile and exhibited personality conflicts. However, they did not find any consistent changes in intellectual ability, psychomotor function, or perceptual ability. This latter finding very likely reflects the differences in individual ability to adapt to nonstimulating conditions.

The stressors that confront isolated groups, whether they be in operational, experimental, or disability-related situations, include boredom, monotony, and the loss of sources of emotional gratification (Mullin, 1960); the lack of privacy and sensory input (Haythorn, 1973); and interpersonal conflicts (Kanas and Fedderson, 1971). The last is influenced by individual adaptability, environmental stressors, and social interaction among the group, including quality of communication, social power or influence, cohesive and disruptive forces present, sociometric structure of the group, status, and leadership and group size.

Interpersonal conflict depends not only on individual adaptability and environmental stressors, but also on the dynamics of group social interaction. Group interaction is influenced by the quality and frequency of communication, social power or influence, cohesive and disruptive forces, individual affect, sociometric structure, status within the group, group size, and leadership.

Results of a 1-year Polish antarctic expedition (Terelak, 1984) noted that the first half of the wintering period was the most difficult in terms of adjustment, as indicated by increased levels of neurotic behavior, anxiety, and introversion (withdrawal). Concomitant changes in activity (work, social interchange, gregariousness, and reflectiveness) were also noted in the second half of the expedition period. The lack of consistency in performance responses to confinement and isolation probably reflects the relative impact of the different environmental, individual, and group factors. Deterioration in performance and mood has been observed in several confinement and isolation studies and includes irritability and hostility (Alluisi *et al.*, 1977), reduced communication and reduced cooperation (Cowan and Strickland, 1965), depression (Strange and Klein, 1973), impaired cognition (Gunderson, 1963), anxiety and headaches (Weybrew, 1962), and reduced alertness, with concentration difficulties (Ritch, 1948). Prolonged confinement in chamber studies designed to simulate spaceflight resulted in decreased psychomotor skill performance, memory, judgment, and learning ability (Chambers, 1964). Intragroup hostility may be displaced onto the outside technical support staff (Fraser, 1966).

Isolation studies of subjects exposed to constant environmental conditions of light, temperature, and humidity, conditions under which there is a greater tendency for the circadian system to desynchronize spontaneously, show somewhat different results. Under such constant conditions, performance improved following the onset of internal circadian desynchronization (Zimmerman *et al.*, 1981). In another series of experiments in isolation in which subjects were exposed to light/dark cycles outside the range of circadian entrainment (20- and 28-hr days), internal circadian desynchronization was associated with improved computation speed and subjective well-being, despite reduced sleep (Wever, 1979). The unexpected improvement in performance in desynchronized subjects was attributed to the young age range of the subjects and the relatively short duration of exposure to isolation (less than 1 month). Under such circumstances where the environment is poor in stimuli, additional stimuli, which originate from disorders of the circadian system, may be advantageous (Wever, 1979).

A number of studies of human performance have been conducted using confined or socially isolated subjects, or both. In a 30-day isolation study, Dushkov and associates (1968) found that calculation performance initially deteriorated, then stabilized after 10 days, and deteriorated again near the end of the isolation period. However, no significant changes occurred in a cognitive test battery during 10 days of confinement with either social interaction or social isolation (Zubek *et al.*, 1969b). Frazer (1966) concluded that confinement of healthy individuals in chamber studies had no significant effect on intellectual or psychomotor function in the absence of sensory or perceptual deprivation. Altman and Haythorn (1967) demonstrated that isolated subjects performed monitoring, decoding, and simulated combat information tasks better than controls who were only confined. Smith (1969), in a review of several confinement studies, concluded that neither intellectual effectiveness nor perceptual ability shows any consistent change in short-term confinement, despite evidence of increased subject irritability, hostility, and personality conflicts between subjects.

The results of longer confinement studies are more ambiguous, since deterioration in motivation seemed to be the only consistent change (Smith, 1969). Interpretation of the results of isolation or confinement studies is confounded by differences in study duration, degree and type of isolation or confinement, type of performance tasks, subject motivation, the nature of social interaction between subjects in group studies, and environmental factors. Altman and Haythorn (1967) interpret the performance inconsistencies observed in confinement studies in terms of an arousal model in which the individual impact of confinement or social interaction factors affect performance by altering levels of arousal. Performance then subsequently increases or decreases, depending on the degree of induced arousal. The increased performance efficiency observed by Wever (1979) and Zimmerman and co-workers (1981) in desynchronized individuals could be interpreted as a response to increased arousal levels consequent to circadian desynchronization rather than to isolation or confinement alone. In fact, Wever (1979) found increases in concentration and decreased fatigue levels in desynchronized subjects.

III. IMMOBILIZATION

A. Psychological Aspects of Bed Rest

Bed rest incorporates not only physical inactivity and physical restriction but also sensory deprivation, which may be one of the most important contributors to psychosocial changes occurring from immobilization. In experimental studies in which subjects are deprived of visual and auditory stimulation, the experience of immobilization is enhanced (Steinberg, 1980), and the restriction of mobility may be a factor in the mental disturbances observed with sensory deprivation (Bexton et al., 1954; Lilly, 1956; Rigs, et al., 1953; Wexler et al., 1958). Ryback and co-workers (1971a) have commented that "in a sense, a familiar and repeatedly performed task, occurrence, or event becomes meaningless neurophysiologically in the same sense that steadily fixated visual test objects disappear." This type of neurophysiological response may underlie the effect or emotional state of monotony and boredom and may be reflected in the slowing of the electroencephalograph (EEG) brain waves observed by Petukhov and Purakhin (1968) in awake subjects during prolonged bed rest (Ryback et al., 1971a,b).

Studies of immobilized patients have shown that sensory deprivation decreases perceptions of patterns and forms, weight discrimination, temperature sensitivity, and speed of perception (Olson, 1967). Time estimation also becomes distorted so that patients tend to confuse the past, present, and future. Time intervals were overestimated by 10 to 13% during a 3 day bed rest (Medvedev and Bagrova, 1977).

Heron (1966) studied subjects bed rested in an enclosure that was lighted 24 hr per day. The subjects were not permitted visual, auditory, or tactile stimulation. This sensory deprivation resulted in significant disturbances in cognitive, perceptual, and motor processes. Some of the subjects stated that during the experience, their thought processes began to change so that they eventually suffered hallucinations related to the senses that had been deprived of stimulation. Investigators have also noted that subjects became irritable, childish, and suspicious toward them (Steinberg, 1980).

Cognitive processes in the absence of stimuli are subject to a number of changes (Heron, 1966; Olson, 1967; Zubek et al., 1969a). Subjects immobilized during learning and motivation studies showed a decrease in "motivation to learn, retain, transfer, and generalize information" (Olson, 1967). Problem-solving became more difficult, and subjects expressed little interest in attempting it. Moreover, they lost the ability to receive the necessary information because of the decrease in sensory stimulation, and their ability to discrimate deteriorated. Kotke (1966) found that progressive dulling of the intellect led to confusion and eventually to a loss of interest in everything, including eating and activity. Hammer and Kenan (1980) in a review of psychological aspects of immobilization concluded that ". . . the person whose intellectual, emotional, physical, or social activities have been limited may be less inclined to initiate activity even when movement is permitted. The process becomes self-intensifying and self-destructive."

Significant psychological and even neurological impairment has been documented in Soviet bed rest studies. Subjects in these studies, ranging up to 120 days in duration, have shown increased fatigue and irritability, emotional lability, weakened memory, increased psychological dependency, and difficulties in logical thinking (Agadzhanyan and Chernyakova, 1982; Bogachenko, 1969; Panferova, 1963; Sorokin et al., 1969; Krupina and Fedorov, 1977; Ioseliani et al., 1985). Bedrested subjects (Agadzhanyan and Chernyakova, 1982) and hospitalized athletes (Izakson, 1978) show reduced cerebral cortical and higher autonomic center activity. Neurological deficits include EEG abnormalities (Yefimenko, 1969) and the development of interhemispheric asymmetry, with functional insufficiency of the dominant hemisphere in most subjects (Panov et al., 1969). In a 62-day bed rest study, Petukhov and Purakhin (1968) found that after 2 or 3 days in bed, awake healthy subjects showed a gradual shift of cortical response toward a slow frequency. Between the 20th and 30th days of bed rest, the subjects showed impairment in central nervous system function, particularly those subjects who had not exercised. These changes were marked by a predominance of slow waves in the EEG, a drop in response rate, and a loss in aural discrimination, all of which indicate dulling of cortical processes.

B. Changes in Sleep

Bed rest studies conducted to evaluate changes in sleep patterns during long-term spaceflight have provided information on clinical immobilization as well. In one study (Ryback *et al.*, 1971a,b), deep sleep increased by 16% in an exercised group during bed rest and by 17% in a nonexercised group; however, the nonexercised group had higher levels of anxiety, hostility, and depression than the exercised group. Time spent in rapid eye movement (REM) sleep increased during bed rest, and this increase persisted during the 6-week recovery period. It is known that bed rest is associated with significant changes in neurohumoral control mechanisms, particularly in pituitary–adrenal function (see Chapter 5). Other factors such as muscle atrophy during bed rest may also play a role by increasing growth hormone secretion during longer periods of deep sleep, with resultant effects (Smith, 1969; Takahashi *et al.*, 1968). The occurrence of non-REM sleep is associated with fatigue (Roffwarg *et al.*, 1966) or with physiological repair mechanisms (Fisher, 1965). Sleep impairment has been commonly observed during bed rest studies (Agadzhanyan and Chernyakova, 1982; Catalano *et al.*, 1977; Panferova, 1963; Krupina and Fedorov, 1977; Ioseliani *et al.*, 1985). These sleep disturbances involve longer sleep latencies, shorter duration of sleep episodes, and more frequent awakenings (Myasnikov, 1975), which appear starting on days 2 or 3 of bed rest (Sorokin *et al.*, 1969) and are present in up to 60% of bed rested individuals (Izakson, 1978). Head-down bed rest may exacerbate the effects of horizontal bed rest by producing headaches, motion sickness symptoms, and visual hallucinations, as well as sleep disturbances (Kakurin *et al.*, 1976). The initial 2–3 days of head down bed rest are commonly associated with proneness to fatigue (Krupina and Fedorov, 1977), decreased mental work capacity (Ioseliani *et al.*, 1985), increased errors in perception of optical coordinates (Bokhov *et al.*, 1975) and impaired sleep (Kakurin *et al.*, 1976). THe degree of deterioration is dependent upon individual personality type (Ioseliani *et al.*, 1985). The initial deterioration in sleep and performance probably results from the postural change and cephalad fluid shifts (Kakurin *et al.*, 1976; Taranenko 1981) since decrements in tracking performance were observed immediately after exposure to postural shifts (Taranenko 1981; Avetikyan *et al.*, 1977). Head down bed rest beyond 5 days is often characterized by increased complaints of fatigue, inability to concentrate, and reported increases in subject irritability and unstable affect (Krupina and Fedorov, 1977; Ioseliani *et al.*, 1985).

C. Decrements in Performance

Zav'yalov and Mel'nik (1967), using a tracking test found a significant deterioration in compensatory scanning in a 10-day bed rest study and a significant deterioration in flight control skill (complex psychomotor) in a 100-day bed rest study (Zav'yalov, *et al.*, 1970). Ryback and associates (1971a), in a 35-day bed rest study, found significant decrements in reaction time and decreased performance in a simple psychomotor task but not in a complex psychomotor task.

However, in a second study of similar duration, no significant changes in performance were found (Ryback et al., 1971b). The difference was attributed to the fact that in the first study the subjects were tested while standing, but in the second study, they were supine. Ryback and associates (1971b) did find a significant decrease in handgrip strength, and this has been reported by others (Greenleaf et al., 1977). After 60 days (Bourne et al., 1968), 10 days (Zav'yalov and Mel'nik, 1967), 7 days (Trimble and Lessard, 1970), and 21 days of bed rest (Taylor et al., 1949), no significant changes in handgrip strength were found. Ryback and associates (1971b) claimed that their finding of a decrease in handgrip strength occurred because they tested the nondominant hand (left) and that somatosensory functions are more diffusely represented in the contralateral right cerebral hemisphere. However, Trimble and Lessard (1970) had also used the nondominant hand. In their 21-day study, Taylor and associates (1949) found a slight decrement in neuromuscular coordination but no change in hand speed or reaction time.

Other investigators have found no change in other performance parameters during bed rest. These studies include psychomotor and problem-solving in a 14-day bed rest study (Storm and Gianetta, 1974); visual response time in another 14-day bed rest study (Haines, 1973); tracking (simulator) in a 10-day bed rest study (Chambers and Vykukal, 1975; Winget et al., 1979); reaction time, psychomotor tests, and memory in a 7-day bed rest study (Trimble and Lessard, 1970); and tracking in a 28-day bed rest study (Meehan et al., 1966). However, in a 17- and 19-day study in which subjects were immobile but not recumbent Dushkov (1967), reported that impaired neuromuscular coordination developed.

The relative effects of recumbency, immobilization, and perceptual deprivation upon performance were examined by Zubek and co-workers in a series of studies. One week of immobilization resulted in decrements in perceptual–motor, intellectual, and memory tests but not in vigilance (Zubek and MacNeill, 1966; Zubek and Wilgosh, 1963). Some of these decrements were statistically significant and others were not, depending on the test. They then found that immobilization plus perceptual deprivation resulted in poorer performance on a battery of intellectual and perceptual–motor tests than did either immobilization or recumbency alone (Zubek et al., 1969a).

Cognitive performance following more prolonged inactivity was evaluated in a 30-day bed rest study (Ioseliani, 1974) in which one group of subjects was exposed to lower body negative pressure (LBNP) for 3 hr twice a day throughout the study in an effort to counteract the detrimental effect of bed rest. A second group was exposed to only fractional doses of LBNP during the last 5 days of bed rest. The LBNP-treated subjects showed a nonuniform deterioration in cognitive performance (counting, memory) and by the 24th day were unable to work because of difficulty in concentrating. The second group experienced insignificant decreases in mental productivity. Therefore, LBNP during bed rest may actually aggravate mental deterioration occurring with bed rest. Spencer and associates (1965), in a study of paralyzed patients, reported that the effect of physical inactivity on the central nervous system led to "decreased attention span, alterations in body image perception, behavior regression, depression, and emotional lability."

Although no two studies have been conducted in exactly the same way, certain generalizations can be drawn from the results of these complex and often contradictory performance studies: (a) performance decrements are more apparent in the longer bed rest studies in which subjects are more strictly confined to bed rest or immobilized, and (b) the choice of performance test utilized is very important in evaluating performance changes. Inconsistencies in the relationship between bed rest and performance changes can be attributed to the following factors: (a) the type of performance test chosen may not have been sensitive to the effects of bed rest; (b) test-to-test variability in performance may mask subtle performance changes; (c) circadian changes or sampling rate aliasing effects may obscure bed-rest-dependent performance effects in studies in which performance is measured once a day; (d) the effect of an independent variable such as bed rest may be more evident during skill acquisition than at performance level asymptotes, at which performance testing occurs (Storm and Gianetta, 1974); (e) presentation of group performance means may obscure significant changes in the performance of individuals; (f) performance decrements may be masked by increasing skill levels through repetition or by increased effort; and (g) performance changes may be brought about or counteracted by changes in extraneous variables (e.g., arousal, fear, motivation, social interactions) of which the investigator is unaware.

D. Stress Effects

The role of inactivity as a physiological stressor has been established in several studies by measuring changes in levels of blood pressure, catecholamines, and corticosteriods. These studies, ranging from 2–62 days in duration, were primarily bed rest studies but also included one water-immersion and two chair-restraint studies. The results, however, have been inconclusive. In some studies, blood pressure, cortisol, and norepinephrine have increased during bed rest (Goldwater et al., 1977; Leach et al., 1973; Mikhaylovskiy et al., 1967; Natelson et al., 1982; Vallbona et al., 1965; Vernikos-Danellis et al., 1974). In others, epinephrine remained unchanged (Blomquist et al., 1971; Chavarri et al., 1977; Chobanian et al., 1974; Maksimov and Domracheva, 1976; Natelson et al., 1982; Rodak et al., 1967; Stevens et al., 1966) or decreased (Balakhovskiy et al., 1972; Buyanov et al., 1968; Cardus et al., 1965; Dlusskaya et al., 1973; Hoffler et al., 1971; Schmid et al., 1971; Vanyushina et al., 1966; Zubek, 1968). In a 62-day study, blood pressure levels varied throughout the investigation (Georgiyevskiy and Mikhaylov, 1968). The inconsistencies in the foregoing studies, therefore, may reflect the existence of nonlinear stressor responses that depend on the length of bed rest. In a more recent study, several physiological parameters were measured every 15 min for 8 hr on bed rest days 1, 3, and 5 (Natelson *et al.*, 1983). None of these variables (e.g., cortisol, catecholamines, body temperature, blood pressure, and heart rate) predicted subsequent proficiency in the subjects' performance with video games. However, principal components analysis suggested that an "alertness" factor—consisting of catecholamine and blood pressure levels and self-rated mood—was a reliable predictor of performance during bed rest.

E. Psychosocial Aspects

Perhaps the greatest deprivation to the immobilized individual or patient is the reduction in, or lack of, social interaction. Such interaction provides stimuli for all of the senses and intellectual and emotional stimulation as well. When socially isolated, humans will very soon begin to talk to inanimate objects or to themselves (Steinberg, 1980).

In clinically bed rested individuals, social interaction can be disturbed either by withdrawal of the individual or by reduced contact with family and friends. The lack of social interaction can lead to decrements in performance and to functional impairment (Hyman, 1972). Litman (1961) found that socially restricted, orthopedically disabled patients were substantially less motivated in rehabilitation programs than more socially active patients. Sussman (1964) claimed that tubercular patients who did not live with or interact with family members were less likely to be rehabilitated medically, vocationally, economically, and socially. Others have also found that lack of social contacts decreased the potential for successful recovery in a variety of health problems (Roth and Eddy, 1967). Hammer and Kenan (1980) have pointed out that an increasing amount of evidence, derived from studies of a variety of populations and treatment programs, demonstrates that the degree of social isolation (before, during, and after treatment) not only can impair rehabilitation but can adversely affect the seeking of treatment. Patients who had someone guiding and protecting them and ensuring that the needed care was received responded more rapidly and effectively to treatment.

In recent studies, short-duration (1–7 days) bed rest has been combined with social and sensory deprivation (Campbell, 1984; Nakagawa, 1980; Ohta, 1983). Subjects were isolated from social contacts, television, reading, and time cues. Total sleep time and sleep latency were measured during bed rest. Sleep episodes became fractioned into about 4-hr (Campbell, 1984; Nakagawa, 1980) or 12-hr ultradian rhythmic patterns (Ohta, 1983). These findings show that periodic diurnal environmental stimuli are necessary to maintain the temporal organization in the sleep/wake cycle during restricted mobility.

F. Chronophysiological Aspects

Circadian rhythms are normally synchronized to a 24-hr day by periodic zeitgebers (synchronizers) in the environment. Daily alterations in the light/dark cycle provide the primary zeitgeber for circadian rhythms in animals (Pittendrigh, 1981). In humans, however, there is controversy over whether light (Czeisler et al., 1981; Morgan et al., 1980) or social interaction (Stepanova, 1973; Vernikos-Danellis and Winget, 1979; Wever, 1979) is the primary zeitgeber. The fact that body temperature rhythm timing follows seasonal changes in sunset time, independent of social routine (Morgan et al., 1980), and that blind individuals often exhibit free-running non-24-hr periods despite exposure to periodic social contacts (Mills et al., 1977) supports the role of the light/dark cycle as the primary zeitgeber. On the other hand, substantial evidence supports the importance of social contact as a primary zeitgeber. Wever (1979) found that some individuals exhibited free-running circadian rhythms in isolation despite exposure to 24-hr light/dark cycles. Furthermore, he found that subjects isolated in groups in constant light tended to synchronize their circadian rhythms and to exhibit longer circadian periods, with reduced period variability, as compared with the same subjects isolated individually under the same light conditions (Wever, 1979). Winget and associates (1974c) demonstrated that two groups of 3 subjects, each of which was exposed to constant light in isolated but identical conditions, showed identical period lengths within a group but significantly different circadian period lengths between groups. This response was attributed to the synchronizing influence of one psychologically dominant individual in each group. Finally, individuals isolated on different photoperiods can synchronize their circadian rhythms as a consequence of verbal communication through an intercom (Forgays, 1973). Therefore, social cues may provide the primary zeitgeber influence in humans, but light/dark cycles may be more effective zeitgebers if the light intensity exceeds 3000 lx, since bright light has a direct physiological effect upon the circadian range of entrainment (Wever, 1983a).

As early as 1873, it was demonstrated that the circadian rhythm of body temperature persists in bed rested or fasting subjects (Jurgenson, 1873). However, the persistence of this rhythm was attributed to subtle diurnal changes in muscular activity or environmental factors (Johansson, 1898). It is now generally accepted that physiological circadian rhythms are generated by a dual endogenous pacemaker system in the central nervous system (Moore-Ede et al., 1982; Wever, 1979). More recent studies have revealed significant alterations in circadian rhythmic stability in male and female subjects during 7–120 days of bed rest. During bed rest, the circadian amplitude or range of oscillations decreases in such variables as oral temperature (Aschoff and Wever, 1980; Lkhagva 1980, 1981; Panferova, 1963), pulse rate (Aschoff and Wever, 1980; Lkhagva, 1981, 1984; Panferova, 1963; Winget et al., 1972, 1974a, 1976, 1979), blood pressure

(Panferova 1963, 1978), and urine volume, potassium and chloride (Lobban and Tredre, 1964). Increased amplitude is found in urine calcium (Moore-Ede *et al.*, 1972), plasma renin (Chavarri *et al.*, 1977), and thyroxine (Winget *et al.*, 1974a). The amplitude of the plasma 17-hydroxycorticosteroid rhythm remains unchanged (Cardus *et al.*, 1965), but the amplitude of the plasma cortisol rhythm may decrease (Erlich *et al.*, 1974; Vernikos-Danellis *et al.*, 1972; Winget, *et al.*, 1977) or remain unchanged (Chavarri, *et al.*, 1977).

Although the amplitude of the oral temperature rhythm decreases during bed rest, the amplitude of continuously sampled rectal temperature does not change during 10–14 days of bed rest (Winget *et al.*, 1976, 1979). Such amplitude decreases are more likely result from decreases in the peak or maximal levels of the physiological variable than to increases in the trough or minimal levels (Aschoff and Wever, 1980). These amplitude changes are also often associated with circadian rhythm phase dissociation, or desynchronization. The rectal temperature rhythm acrophase generally occurs 1 hr later than that of heart rate during bed rest (Winget *et al.*, 1976, 1979). Extensive rhythm dissocation has been observed only occasionally in a single individual during 10 days of bed rest (Winget *et al.*, 1976) but is more prevalent at 22–24 days of bed rest (Winget *et al.*, 1974b). An 8-hr shift in the peak time of the heart rate rhythm with respect to body temperature was observed during bed rest days 10 through 20 and was associated with an inversion of the urine volume and electrolyte rhythm peak times (Agadzhanyan and Chernyakova, 1982). Reversal of the phase of insulin rhythm has also been observed during bed rest (Vernikos-Danellis *et al.*, 1976).

Circadian rhythm waveforms during bed rest are often distorted (Volkov, 1979) or flattened (Panferova, 1963) or become more sinusoidal (Winget *et al.*, 1976). A transition from a unimodal 24-hr pattern to a bimodal 12-hr pattern has been found for body temperature (Panferova, 1963; Volkov, 1979), locomotor activity (Catalano *et al.*, 1977), and sleep/wake cycles (Ohta, 1983) during bed rest. Other investigators (Hildebrandt *et al.*, 1974) have associated this transposition in rhythm frequency with the development of fatigue states.

There are several other interesting chronophysiological aspects of immobilization in humans. Anti-orthostatic bed rest induces greater changes in circadian rhythm indices than in horizontal bed rest (Agadzhanyan and Chernayakova, 1982; DeRoshia and Winget, 1979). Circadian amplitude and phase instability increase with bed rest duration (Agadzhanyan and Chernyakova, 1982). In a 120-day bed rest study, Baevskiy and co-workers (1977) described and characterized the phases of circadian rhythm response as follows: (a) the "stress" phase at 60 days, in which maximum rhythm amplitudes were observed; (b) the "overstress" phase at 90 days, at which rhythm amplitudes were relatively low but rhythm synchronization increased; and (c) the "asthenization" phase, in which changes in circadian indices were inconsistent and rhythm synchronization was lost.

During bed rest, the body temperature amplitude is higher in subjects covered by blankets than in uncovered subjects (Aschoff and Wever, 1980). Changes in body temperature levels during bed rest, therefore, may result from impaired thermoregulation (Panferova, 1978; Williams and Reece, 1972), but it is not clear whether the change in thermoregulation results from inactivity or from changing phase relationships between rhythmic components of the thermoregulatory system (Fuller *et al.*, 1978). Finally, the transition between bed rest and post-bed-rest recovery is more difficult, as measured by the disruption of circadian rhythm indices (e.g., mean level, amplitude), than the transition from pre-bed-rest to bed rest (Winget *et al.*, 1979). This difference probably reflects increased orthostatic intolerance during post-bed-rest recovery as a consequence of the progressive physiological deconditioning induced by bed rest.

Deterioration in circadian rhythm stability during bed rest has been attributed to (a) neuropsychological dysfunction, (b) the physiological effects of bed rest, and (c) reduced effectiveness of cyclic environmental synchronizers.

The first hypothesis presumes that circadian dissociation results from neuro-psychological changes including increased anxiety, hostility, and depression; emotional lability, and neurological abnormalities. Other studies have shown that behavioral stress can produce rhythm desynchronization (Stroebel, 1969) and that individuals with a greater tendency to neurotic behavior show a greater degree of internal rhythmic desynchronization during isolation in constant light (Lund, 1974). This hypothesis, then, predicts that the circadian dissociation observed during bed rest results directly from the neuropsychological disturbances that are a direct consequence of bed rest.

The second hypothesis presumes that circadian rhythm dissociation results from physical inactivity during bed rest. Postural changes involved in bed rest, hydrostatic pressure changes, reduced afferent proprioceptor input, or redistribution of body fluids and electrolytes may be primarily responsible for rhythm instability (Koroleva-Munts, 1974; Vernikos-Danellis *et al.*, 1974). This view is based on the primary dependency on different synchronizers for the regulation of various physiological rhythms. Certain physiological variables, such as aldosterone (Bartter *et al.*, 1972) and the thyroid hormones, may be posture-dependent, but others, such as cortisol, appear to be controlled by environmental or social synchronizers (Vernikos-Danellis *et al.*, 1974). However, it is not clear how reduced proprioceptor input per se is involved in the rhythm destabilization process.

The third hypothesis presumes that circadian dissociation during bed rest results from reduced overall effectiveness of environnmental synchronizers subsequent to the elimination of the cyclic alteration between activity and rest. As previously mentioned, bed rest can be considered to be a form of sensory deprivation in which social interaction becomes an increasingly important influence. Bed rest also eliminates the regular diurnal alteration between rest and activity and between the transverse and horizontal application of gravitational force to the upright or supine body, respectively. In this sense, bed rest may reduce the relative strength of the environmental synchronizers or may eliminate the diurnal changes in activity and hydrostatic forces which reinforce the synchronizing influence of the light/dark cycle. Soviet investigators (Alyakrinskiy, 1980; Mikushkin, 1969) have emphasized the importance of maintaining a well-defined 24-hr activity/rest cycle to facilitate the maintenance of circadian rhythm synchronization. Circadian rhythmicity remains synchronized in constant conditions when at least 4-hr of sleep time take place at the same time each day (Minors and Waterhouse, 1981). However, this "anchor sleep" is apparently too weak a cue to prevent circadian dissociation during bed rest. Phase shifts during bed rest may result from interactions between dual oscillatory pacemaker systems underlying the heart rate and body temperature circadian rhythms (Kronauer et al., 1982). The two oscillatory pacemaker systems would be differentially sensitive to the reduced effectiveness of the light/dark cycle and the absence of regular alterations in rest and activity.

Because of the complex interrelationships between the effects of inactivity, circadian rhythm dissociation, and neuropsychological function, it is possible that one or more of these hypotheses may account for the observed circadian dissociation during bed rest.

Social factors, as well as the light/dark cycle, may play a role in maintaining rhythm synchronization during bed rest as they do under ambulatory conditions. In a 7-day bed rest study, one group of 4 subjects housed in the same room showed a significantly earlier time of body temperature rhythm acrophase and lower group acrophase variance during the control period than did another group of subjects who slept in another room but otherwise were maintained under identical conditions. The subjects in each group were selected so that the levels of several physiological parameters would be equivalent for each group. The occurrence of earlier acrophase times in the first group was attributed to the presence of a socially dominant individual with the most extreme morningness chronotype (DeRoshia and Winget, 1979).

Although significant differences exist between· male and female circadian rhythm indices (e.g., mean level, amplitude, period length) in ambulatory subjects (Winget et al., 1977), we have observed no differences in circadian rhythm dissociation between male and female subjects during bed rest. In female subjects, however, a menstrual cycle influence is present during bed rest. Body temperature is 0.3°C higher in postovulatory than in preovulatory subjects, and there is a 0.1°C change in the body temperature rhythm amplitude between these two phases of the menstrual cycle (Aschoff and Wever, 1980).

Exercise regimens and periodic exposure to hypobaric gas mixtures have been studied as countermeasures to bed-rest-induced rhythm disturbances. Hypobaric gas mixtures reportedly reversed the phase shift and reduction in amplitude of the body temperature rhythm in bed rested subjects as compared with control subjects (Agadzhanyan and Chernyakova, 1982). A moderately heavy exercise regime also prevented the phase shift in heart rate during head-down bed rest and reversed the phase shift observed in the body temperature rhythm during horizontal bed rest (Agadzhanyan and Chernyakova, 1982). However, in another study, exercise was not effective in preventing bed-rest-induced rhythm alterations (Vernikos-Danellis et al., 1972).

Several circadian rhythm indices have been identified that may be useful in predicting individual adaptability to bed rest. In a bed rest study with female subjects, there was a significant negative correlation between the pre-bed-rest body temperature rhythm amplitude and the degree of temperature rhythm phase shift during bed rest (Winget et al., 1979). The results indicated that individuals with relatively low rhythm amplitude have a greater tendency for rhythm dissociation during bed rest. In another female bed rest study, mean daily heart rate levels and the amplitude of the plasma cortisol rhythms before bed rest correlated significantly with centrifuge acceleration tolerance after 14 days of bed rest (Vernikos-Danellis et al., 1978; Winget et al., 1976).

G. Rhythm Desynchronization, Sleep Disturbance, and Performance

The previous discussion has provided substantial evidence that circadian rhythm dissociation or instability and sleep disturbances frequently occur in bed rested subjects or patients, particularly in the absence of environmental stimuli. Since these changes in the temporal organization of physiological processes are often associated with deterioration in performance and well-being, it is important to examine the relationship between altered circadian rhythmicity and behavioral function.

A large body of literature documents significant deterioration in performance in response to sleep loss or sleep disturbance (Beljan et al., 1972; Holley et al., 1981; Johnson and Naitoh, 1974). Performance deterioration associated with sleep loss may result from lapses of vigilance rather than a general decline in performance (Webb, 1968) or from lowered levels of arousal (Broadbent, 1963). The more complex and difficult psychomotor tasks and long-duration tasks are the ones most sensitive to sleep loss (Naitoh, 1969). Sleep disturbance appears to be more important than sleep loss in performance deterioration (Johnson and Naitoh, 1974; Nicholson, 1970), perhaps by potentiating the circadian influence on performance or by augmenting the effects of other stressors. A significant deterioration occurs in vigilance and calculation performance following 2- to 4-hr advances in the sleep/wake cycle, despite the maintenance of normal sleep duration (Taub and Berger, 1974). The results of Taub and Berger indicate that performance deterioration may result from circadian rhythm disturbance and not solely from sleep loss.

The effects of alterations in timing of work/rest schedules or environmental synchronizers upon subsequent performance and mood have been documented in numerous ground-based experimental designs (e.g., shift work, transmeridian flight, altered day length, and shifted light/dark cycle experiments). The results of these studies show that significant decrements in performance, sleep disturbance, and increased psychosomatic complaints result when the timing of work/rest schedules or light/dark cycles is advanced (Holley et al., 1981; Klein and Wegmann, 1980; Litsov, 1979; Wright et al., 1983).

Performance levels are more adversely affected following phase advances (simulated eastward flight) than following phase delays (simulated westward flights) in ground-based, phase-shift studies. Wever (1980) found significant impairment in psychomotor performance after 6-hr phase advances of the light/dark cycle but not after phase delays. When the sleep/wake cycle was advanced 1.5 hr every 5 days, subjects exhibited increased sleep disorders and psychomotor and calculation performance decrements (Litsov, 1979). Significant deterioration in vigilance, calculation proficiency, and mood was observed in subjects whose sleep/wake period was advanced 2-4 hr (Taub and Berger, 1974). Female subjects advanced by 8 hr exhibited a 17–38% deterioration in short-term memory, reaction time, and visual search performance (Preston, 1978). Student pilots whose sleep/wake periods were advanced 2.5 hr exhibited sleep impairment, a 35% increase in flight performance errors, and a 14–15% decrement in a standard letter cancellation performance test (Yurchenko, 1981).

Studies have indicated that sleep onset and duration are dependent upon the phase of the body temperature rhythm (Czeisler et al., 1980; Zulley et al., 1981). Deviation from this narrow range of internal circadian rhythm phase relationships results in dissociation between various circadian rhythm systems (dysrhythmia). Dysrhythmia in humans follows advances in either light/dark cycle or work/rest schedules and most likely accounts for the subsequently observed fatigue and sleep disturbances (Beljan et al., 1972; Holley et al., 1981).

There is relatively high individual variability in the deterioration of human performance and the subsequent time required to readapt in response to circadian dysrhythmia. For example, although 25–30% of transmeridian travelers have little or no difficulty adjusting to jet lag, an equal percentage adjust poorly or not at all (Klein and Wegmann, 1980). Factors such as rhythm stability (Stepanova, 1977), rhythm amplitude (Andlauer and Reinberg, 1979), rhythm period length (Wever, 1983b, 1984b), behavioral traits (Colquhoun, 1979, 1984), chronotype (morningness/ eveningness, Colquhoun, 1979), age (Hauty and Adams, 1965), motivation (Graeber, 1982), and sleep habits (Folkard and Monk, 1979) have been found to influence an individual's capacity to adjust to induced circadian dysrhythmia.

It is clear that circadian dysrhythmia induced by environmental factors results in significant deterioration in performance and well-being in susceptible individuals. However, the previously mentioned studies of human subjects isolated in constant environmental conditions showed that performance and psychological well-being may actually improve under these conditions, despite the occurrence of internal rhythmic desynchronization in these subjects. The apparent differences between these study results most likely depend on whether dysrhythmia occurs spontaneously, as in the constant environment experiments, or as the result of alterations in the timing of environmental or social synchronizers ("forced desynchronization"). It is not evident whether the circadian rhythm disturbances observed during bed rest studies represent spontaneous dissociation or forced desynchronization (e.g., by inactivity or neuropsychological changes). In the latter case, chronic bed rest or weightlessness during space missions may make it difficult to maintain human performance, sleep, and well-being without appropriate countermeasures.

IV. SUMMARY AND CONCLUSIONS

A. Experimental Aspects

The occurrence of deterioration in mood and performance in isolated and confined humans depends on individual personality factors; group size, structure, and interactions; environmental stressors; and the duration of isolation. Bed rest incorporates a degree of sensory deprivation, isolation, and confinement. Prolonged bed rest or immobilization may result in sleep disturbances, neurological changes, increased emotional lability and fatigue, and deterioraton in motivation and cognitive performance. Deterioration in other performance tasks is more likely to occur in bed rest of longer duration, particularly if a high degree of immobilization is maintained. The negative effects of bed rest on neuropsychological function are compounded by physical stress, sensory deprivation, and increasing duration of bed rest. Circadian rhythmicity deteriorates during bed rest, as indicated by a reduced amplitude in most rhythms, rhythmic phase dissociation, and a tendency to shift from 24-hr to 12-hr or bimodal rhythmic patterns, which are indicative of fatigue states. This deterioration increases with bed rest duration and is greater in head-down than in horizontal bed rest. Circadian rhythmic instability during bed rest may result from neuropsychological dysfunction, immobilization itself, or reduced effectiveness of environmental synchronizers. This instability may be reversed by social interaction, hypobaric gas mixtures, or exercise regimens. Both light/dark cycles and social contact are involved in the synchronization of circadian rhythms. Shifts in the timing of environmental synchronizers can induce circadian rhythm dissociation and subsequent impairment in performance, sleep quality, and psychological well-being. The most important factors that generate dysrhythmia are advances in the timing of environmental synchronizers and forced rather than spontaneous rhythm desynchronization. The degree of circadian dysrhythmia is dependent upon individual characteristics. Circadian rhythmic disturbances may have negative consequences for human performance, sleep, and well-being during chronic bed rest or space missions. Furthermore, circadian rhythmic oscillations and superimposed altered levels of physiological processes during bed rest or spaceflight may potentially push the levels of these processes beyond the limits at which homeostatic feedback control operates. More research is needed to selectively isolate the relative impact of environmental and social stimuli, altered rest/activity schedules, isolation, and immobilization on human health and performance and to better determine the interrelationships between these variables.

B. Clinical Aspects

Several measures can be taken that may be helpful in countering the physiological and psychological deterioration that occurs during bed rest or immobilization. Bed rested patients should be exposed to regular light/dark cycles and periodic daily social contacts. Ambient light intensity should be relatively high (i.e., 2000 lux or more), since bright light often results in the remission of depression (Kripke et al., 1983; Rosenthal et al., 1984), possibly by reinforcing optimal phase relationships between the weaker and stronger circadian oscillatory components (Kripke et al., 1983). Patients should also be exposed to daily sensory stimuli (e.g., television, radio, reading material) to reinforce the normal temporal organization of sleep and to reduce boredom. Social interaction with friends, family, staff, and other patients should be encouraged, since it contributes to the emotional well-being of the patient and reinforces circadian rhythm synchronization. Negative social interaction (e.g., personal conflicts) should be avoided, since it may reinforce psychological deterioration in the bed rested patient, who is confined and isolated from normal social contacts. Daily exercise regimens should be used to faciliate circadian rhythm synchronization, to improve sleep quality, and to reduce the neuropsychological deterioration that results from prolonged inactivity.

REFERENCES

Agadzhanyan, N.A. and Chernyakova, V.N. (1982). *Fiziologia Cheloveka*. **8**, 179-191.
Alluisi, E.A., Coates, G.D., Morgan, B.B. (1977). *In*: "Vigilance", (R.R. Mackie, ed.), pp. 361-421. Plenum Press, New York.
Altman, I. and Haythorn, W.W. (1967). *Human Relations* **4**, 313-340.
Alyakrinskiy, B.B. (1980). *Kosm. Biol. Med.* **14(1)**, 3-8.

Andlauer, P. and Reinberg, A. (1979). *Chronobiol.* **6** *(Suppl. 1)*, 67-73.

Aschoff, J. and Wever, R. (1980). *Klin. Wochenschr.* **58**, 323-335.

Avetikyan, Sh.T., Zingermann, A.M., and Menitskiy, D.N. (1977). *Fiziologia Cheloveka* **3**, 678-684.

Baevskiy, R.M., Niulina, G.A., and Semenova, T.D. (1977). *Fiziologia Cheloveka* **3**, 387-392.

Baker, T.C. and Rohrer, J.H. (1960). Symposium on Human Problems in the Utilization of Fallout Shelters. National Academy of Sciences, Washington, DC (Disaster Study No. 12).

Balakhovskiy, I.S., Bakhteyeva, V.T., Beleda, R.V., Biryukov, Ye.I., Vinogradova, L.A., Grigor'yev, A.I., Zakharova, S.I., Dlusskaya, I.G., Kiselev, R.K., Kislovskaya, T.A., Kozyrevskaya, G.I., Noskov, V.B., Orlova, T.A., and Sokolova, M.M. (1972). *Kosm. Biol. Med.* **6(4)**, 68-72.

Bartter, F.C., Delea, C.S., and Halberg, F. (1972). *Ann. N.Y. Acad. Sci.* **98**, 969-983.

Beljan, J.R., Rosenblatt, L.S., Hetherington, N.W., Lyman, J.L., Flaim, S.T., Dale, G.T., and Holley, D.C. (1972). "Human Performance in the Aviation Environment", NASA Report NAS2-6657, Part I-A.

Bexton, W.H., Heron, W., and Scott, T.H. (1954). *Canada J. Physiol.* **8**, 70-76.

Blomquist, G., J.H. Mitchell, and B. Saltin (1971). *In*: "Hypogravic and Hypodynamic Environments" (R.H. Murray and M. McCally, eds.), NASA SP-269, pp. 171-185. Washington, DC.

Bogachenko, V.P. (1969). *In*: "Problems of Space Biology, Vol. 13, Prolonged Limitations of Mobility and Its Influence on the Human Organism", (A.M. Genin and P.A. Sorokin, eds.), NASA TT-F-639, pp. 170-174, Nauka Press, Moscow.

Bokhov, B.B., Kornilova, L.N., and Yakoleva, I. Ya. (1975). *Kosm. Biol. Aviakosm. Med.* **9(1)**, 51-56.

Bourne, G.H. Nandy, K., and Golarz de Bourne, M.N. (1968). *In*: "Hypodynamics and Hypogravics" (R.H. Murray and M. McCally, eds.), pp. 197-212, Academic Press, New York.

Broadbent, D.E. (1963). *Q. J. Expt. Psychol.* **15**, 205-211.

Buyanov, P.V., Beregovkin, A.V., and Pisarenko, N.V. (1968). *Kosm. Biol. Med.* **1(1)**, 78-82.

Calvin, M. and Gazenko, O.G. (1975). "Foundations of Space Biology and Medicine, Vol. II. Ecological and Physiological Bases of Space Biology and Medicine", NASA Scientific and Technical Information Office, Washington, DC.

Campbell, S.S. (1984). *Psychophysiol.* **24**, 106-113.

Cardus, D.C., Vallbona, C., Vogt, F.B., Spencer, W.A., Lipscomb, H.S., and Eik-nes, K.B. (1965). *Aerosp. Med.* **36**, 524-528.

Catalano, G.T., Winget,C.M., Laursen, A., Sandler, H., DeRoshia, C.W., and Rietman, J. (1977). *In*: "Proceedings, San Diego Biomedical Symposium, Vol. 16", (J.I. Martin, ed.), pp. 375-385, Academic Press, New York.

Chambers, R.M. (1964). *In*: "Physiological Problems in Space Exploration" (J.D. Hardy, ed.), pp. 231-292, Charles C. Thomas Publishers, Springfield, IL.

Chambers, R.M. and Fried, R. (1963). *In*: "Physiology of Man in Space" (J.H. Brown, ed.), pp. 173-256, Academic Press, New York.

Chambers, A.B. and Vykukal, H.C. (1975). *Aerosp. Med. Assoc. Preprints*, pp. 130-140.

Chavarri, M., Ganguly, A., Leutscher, J.A., and Zager, P.G. (1977). *Aviat. Space Environ. Med.* **48**, 633-636.

Chobanian, A.V., Lille, R.D., Tercyak, A., and Blevins, P. (1974). *Circulation* **49**, 551-559.

Colquhoun, W.P. (1979). *Int. Arch. Occup. Environ. Health* **42**, 149-157.

Colquhoun, W.P. (1984). *Aviat. Space Environ. Med.* **55**, 493-496.

Cowan, T.A. and Strickland, D.A. (1965). The Legal Structure of a Confined Microsociety (A report on the cases of Penthouse II and III). Internal working paper No. 34, Space Sciences Laboratory, Social Sciences Project, University of California, Berkeley, CA.

Czeisler, C.A. (1978). Human Circadian Physiology: Internal Organization of Temperature Sleep-Wake and Neuroendocrine Rhythms Monitored in an Environment Free of Time Cues. Ph.D. Thesis, Stanford University, Stanford, CA.

Czeisler, C.A., Weitzman, E.D., Moore-Ede, M.C., Zimmerman, J.C., and Knauer, R.S. (1980). *Science* **210**, 1264-1267.

Czeisler, C.A., Richardson, G.S., Zimmerman, J.C., Moore-Ede, M.C., and Weitzman, E.D. (1981). *Photochem. Photobiol.* **34**, 239-247.

DeRoshia, C.W. and Winget, C.M. (1979). *In*: "Preliminary Report of the Joint US/USSR Hypokinesia Study", 6 p., NASA-Ames Research Center, Moffett Field, CA.

Dlusskaya, I.G., Vinogradov, L.A., Noskov, V.B., and I.S. Balakhovskiy, I.S. (1973). *Kosm. Biol. Med.* **7(3)**, 43-48.

Dushkov, B.A. (1967). *Kosm. Biol. Med.* **1(2)**, 64-70.

Dushkov, B.A., Zolotukhin, A.N., and Kosmolinskiy, F.P. (1968). *Kosm. Biol. Med.* **2(2)**, 64-70.

Edholm, O.G. and Gunderson, E.K.E. (1973). "Polar Human Biology", Heinemann Medical Books, London.

Eilbert, L.R. and Glaser, R. (1959). *J. Appl. Psych.* **13**, 271-274.

Epstein, E.N. (1962). "Report No. 384, Prediction of Adjustment to Prolonged Submergence Aboard a Fleet Ballistic Missile Submarine: II Background Variables", U.S. Naval Medical Research Laboratory.

Erlich, S., Weitzman, E.D., and McGregor, P. (1974). *Sleep Res.* **3**, 168.

Fisher, C. (1965). *J. Am. Psychoanal. Assn.* **13**, 197-270.

Folkard, S. and Monk T.H. (1979). *Ergonomics* **22**, 79-91.

Forgays, D.G. (1973). "Isolation and Sensory Communication", Air Force Office of Scientific Research Project Final Report AD-777156, 80 p.

Fraser, T.M. (1966). *In*: "The Effects of Confinement as a Factor in Manned Space Flight", NASA Report CR-511, Washington, DC.

Fuller, C.A., Sulzman, F.M., and Moore-Ede, M.C. (1978). *Science* **199**, 794-795.

Georgiyevskiy, V.S. and Mikhaylov, V.M. (1968). *Kosm. Biol. Med.* **2(3)**, 48-51.

Goldwater, D., Sandler, H., Rositano, S., and McCutcheon, E.P. (1977). *Aerosp. Med. Assoc. Preprints*, pp. 240-241.

Graeber, R.C. (1982). *In*: "Rhythmic Aspects of Behavior", (F.M. Brown and R.C. Graeber, eds.), pp. 173-212, Erlbaum Assoc., Hillsdale, New Jersey.

Greenleaf, J.E., Bernauer, E.M., Juhos, L.T., Young, H.L., Morse, J.T., and Staley, R.W. (1977). *J. Appl. Physiol.* **43**, 126-132.

Gunderson, E.K.E. (1963). *Arch. Gen. Psych.* **9**, 362-368.

Gunderson, E.K.E. (1974). "Human Adaptability to Antarctic Conditions", American Geophysical Union, Washington, DC.

Gunderson, E.K.E., Nelson, P.D., and Orvick, J.M. (1964). "Personal History Correlates of Military Performance at a Large Antarctic Station", Report No. 64-22, U.S. Navy Neuropsychiatric Research Unit.

Haines, R.F. (1973). *Aerosp. Med.* **44**, 425-432.

Hammer, R.L. and Kenan, E.H. (1980). *In*: "The Immobilized Patient: Functional Pathology and Management", (F.U. Steinberg, ed.), pp. 123-149, Plenum Medical Book Company, New York.

Hammes, J.A. (1965). "Shelter Occupancy Studies, Final Report AD-635-501", University of Georgia.

Hauty, G.T. and Adams, T. (1965). *In*: "Circadian Clocks", (J. Aschoff, ed.), pp. 413-425, North-Holland, Amsterdam.

Haythorn, W.W. (1973). *In*: "Man in Isolation and Confinement", (J.E. Rasmussen, ed.), pp. 218-239, Aldine Publishing Co., Chicago, IL.

Helmreich, R.L. (1966). "Prolonged Stress in Sea Lab II: A Field Study of Individual and Group Reactions", Unpublished Doctoral Dissertation, Yale University.

Helmreich, R.L. (1971). *In*: "Tektite 2: Scientists-in-the-Sea", (R.A. Walter, ed.), U.S. Department of the Interior, pp. VIII-15-VIII-61, Washington, DC.

Heron, W. (1966). *In*: "Frontiers of Psychosocial Research", (D. Coppersmith, ed.), Freeman Press, San Francisco, CA.

Hildebrandt, G., Rohmert, W., and Rutenfranz, J. (1974). *Intern. J. Chronobiol.* 2, 175-180.

Hoffler, G.W., Wolthuis, R.A., and Johnson, R.L. (1971). *Aerosp. Med. Assoc. Preprints*, pp. 174-175.

Holley, D.C., Winget, C.M., DeRoshia, C.W., Heinold, M.P., Edgar, D.M., Kinney, N.E., Langston, S.E., Markley, C.L., and Anthony, J.A. (1981). "Effects of Circadian Rhythm Phase Alteration on Physiological and Psychological Variables: Implication to Pilot Performance", NASA TM-81277.

Honigfeld, A.R. (1965). "Technical Memorandum, Group Behavior in Confinement: Review and Annotated Bibliography", pp. 14-65, Human Engineering Laboratories, U.S. Army, Aberdeen Proving Grounds, MD.

Hull, J. and Zubek J.P. (1962). *Percept. Motor Skills* 14, 231-240.

Hyman, D.D. (1972). *J. Chron. Dis.* 25, 85.

Ioseliani, K.K. (1974). *Kosm. Biol. Aviakosm. Med.* 8(4), 134-135.

Ioseliani, K.K., Narinskaya, A.L., and Khisambeyev, Sh. R. (1985). *Kosm. Biol. Aviakosm. Med.* 19(1), 19-24.

Izakson, K.A. (1978). *Voprosy Kurortologii, Fizioterapil Lechebnoy Fiziocheskoy Kul'tury* 4, 81.

Johansson, J.E. (1898). *Scand. Arch. Physiol.* 8, 85-142.

Johnson, L.C. and Naitoh, P. (1974). "The Operational Consequences of Sleep Deprivation and Sleep Deficit", NATO: Advisory Group for Aerospace Research and Development, *AGARD-AG-193*, 44 p.

Jurgensen, T. (1873). *Die Körperwärme des Gesunden Menschen*, Vogel, Leipzig.

Kakurin, L.I., Lobachik, V.I., Mikhailov, M.M., and Senkevich, Y.A. (1976). *Aviat. Space Environ. Med.* 47, 1083-1086.

Kanas, N.A. and Fedderson, W.E. (1971). "Behavioral, Psychiatric, and Sociological Problems of Long-Duration Space Missions", NASA TM X-58067.

Klein, K.E. and Wegmann, H.M. (1980). "Significance of Circadian Rhythms in Aerospace Operations", NATO: Advisory Group for Aerospace Research and Development, AGARDograph No. 247, Technical Editing and Reproduction, London, pp. 64.

Koroleva-Munts, V.M. (1974). *Fiziol. Zhurn.* 40, 1145-1149.

Kotke, F.J. (1966). *J. Am. Med. Assoc.* 196, 825.

Kripke, D.F., Risch, S.C., and Janowsky, D. (1983). *Psychiat. Res.* 10, 105-112.

Kronauer, R.E., Czeisler, C.A., Pilato, S.F., Moore-Ede, M.C., and Weitzman, E.D. (1982). *Am. J. Physiol.* 242, R3-R17.

Krupina, T.N., and Fedorov, B.M. (1977). *Fiziologia Cheloveka* 3, 997-1005.

Leach, C.S., Hulley, S.B., Rambaut, P.C., and Dietlein, L.F. (1973). *Space Life Sci.* 4, 415-423.

Lilly, J.C. (1956). *Psychiatr. Res. Rep.* 5, 1.

Litman, T.J. (1961). "The Influence of Concept of Self and Life Orientation Factors upon the Rehabilitation of Orthopedic Patients", Ph.D. Dissertation, University of Minnesota.

Litsov, A.N. (1979). *Kosm. Biol. Aviakosm. Med.* 13(1), 53-58.

Lkhagva, L. (1980). *Kosm. Biol. Aviakosm. Med.* 14(4), 59-61.

Lkhagva, L. (1981). *Kosm. Biol. Aviakosm. Med.* 15(6), 63-65.

Lkhagva, L. (1984). *Kosm. Biol. Aviakosm. Med.* 18(2), 60-63.

Lobban, M.C. and Tredre, B.F. (1964). *J. Physiol.* 171, 26P-27P.

Lund, R. (1974). *Psychosomat. Med.* 36, 224-228.

Maksimov, D.G., and Domracheva, M.V. (1976). *Kosm. Biol. Aviakosm. Med.* **10(5)**, 52-57.

May, C.B. (1970). "The Man-Related Activities of the Gulf Stream Drift Mission", NASA TM X-64548.

Medvedov, V.I., and Bagrova, N.D. (1977). *Fiziologia Cheloveka* **3**, 288-294.

Meehan, J.P., Henry, J.P., Brunjes, S., and de Vries, H. (1966). "Investigation to Determine the Effects of Long-Term Bedrest on G-tolerance and on Psychomotor Performance." NASA Report CR-62073.

Mikhaylovskiy, G.P., Benevolenskaya, T.V., Petrova, T.A., Yakovleva, I.Ya., Boykova, O.I., Kuz'min, M.P., Savilov, A.A., and Solov'yeva, S.N. (1967). *Kosm. Biol. Med.* **1(5)**, 57-60.

Mikushkin, G.K. (1969). *Kosm. Biol. Med.* **3(1)**, 48-60.

Mills, J.N., Minors, D.S., and Waterhouse, J.M. (1977). *J. Physiol. (London)* **268**, 803-826.

Minors, D.S. and Waterhouse, J.M. (1981). *Intern. J. Chronobiol.* **7**, 165-168.

Moore-Ede, M.C., Faulkner, M.H., and Tredre, B.E. (1972). *Clin. Sci.* **42**, 433-445.

Moore-Ede, M.C., Sulzman, F.M., and Fuller, C.A. (1982). "The Clocks that Time Us: Physiology of the Circadian Timing System" pp. 448, Harvard University Press, Cambridge, MA.

Morgan, R., Minors, D.S., and Waterhouse, J.M. (1980). *Chronobiology* **7**, 331-335.

Mullin, C.S. (1960). *Am. J. Psych.* **117**, 323-325.

Murray, R.H. and McCally, M. (Eds.) (1971). "Hypogravic and Hypodynamic Environments", Washington, DC. NASA SP-269.

Myasnikov, V.I. (1975). *Aviat. Space Environ. Med.* **46**, 401-408.

Myers, T.I., Murphy, D.B., Smith, S., and Goffard, S.J. (1966). "Experimental Studies of Sensory Deprivation and Social Isolation", Technical Report No. 66-8, Human Resources Research Office, George Washington University.

Naitoh, P. (1969). Sleep Loss and Its Effects on Performance. Navy Medical Report No. 68-3, Neuropsychiatric Research Unit, San Diego, CA.

Nakagawa, Y. (1980). *Electroenceph. Clin. Neurophysiol.* **49**, 524-537.

Natelson, B.H., DeRoshia, C., and Levin, B.E. (1982). *The Lancet I (8262)*, 51.

Natelson, B.H., DeRoshia, C.W., Adamus, J., Finnegan, M.B., and Levin, B.E. (1983). *Pavl. J. Biol. Sci.* **18**, 161-168.

Nicholson, A.N. (1970). *Aerosp. Med.* **41**, 626-632.

Nicogossian, A.E., and Parker, J.F., Jr. (1982). "Space Physiology and Medicine", NASA SP-447, Washington, DC.

Ohta, T. (1983). *Psychiatria et Neurologia Japonica (Toyko)* **85(5)**, 302-330.

Olson, E.V. (1967). *Am. J. Nurs.* **67**, 794.

Ormiston, D.W. (1961). "A Methodological Study of Confinement", Wright Patterson AFB Report WADD-TR-61-258, Ohio.

Panferova, N.Y. (1963). *Fiziol. Zhurn. SSSR* **50**, 741-749.

Panferova, N.Y. (1978). *Fiziologia Cheloveka* **4**, 835-839.

Panov, A.G., Lobzin, V.S., and Belyankin, V.A. (1969). In "Problems of Space Biology, Vol. 13" (A.M. Genin and P.A. Sorokin, eds.), Nauka Press, Moscow, NASA TT F-639, pp. 133-147.

Petukhov, B.N. and Purakhin, Yu.N. (1968). *Kosm. Biol. Med.* **2(5)**, 56-61.

Pittendrigh, C.S. (1981). In "Handbook of Behavioral Neurobiology, Vol. 4, Biological Rhythms" (J. Aschoff, ed.), pp. 95-124, Plenum Press, New Jersey.

Preston, F.S. (1978). *J. Psychosomatic Res.* **22**, 377-383.

Rasmussen, J.E. (1973). "Man in Isolation and Confinement." Aldine Publishing Co., Chicago, IL.

Rigs, L.A., Ratliff, F., Cornsweet, J.C., and Cornsweet T.N. (1953). *J. Optical Soc. Am.* **43**, 495-501.

Ritch, T.G. (1948). "Report on the Effects of Prolonged Snorkelling on the Health of Officer and Men and on the General Habitability of the Guppy-Snorkel Submarine U.S.S. Trumpetfish (SS 425)." Report No. 202, USN Submarine Medical Research Laboratory.

Rodak, K., Birkhead, N.C., Blizzard, J.J., Issekutz, B., and Pruett, E.D.R. (1967). *In* Nutrition and Physical Activity, Symposium of the Swedish Nutrition Foundation (V.G. Blix, ed.), pp. 107-113, Almquist and Wikseles, Uppsala.

Roffwarg, H.P., Muzio, J.N., and Dement, W.C. (1966). *Science* **152**, 604-619.

Rosenthal, N.E., Sack, D.A., Gillin, C., Lewy, A.J., Goodwin, F.K., Davenport, Y., Mueller, P.S., Newsome, D.A., and Wehr, T.A. (1984). *Arch. Gen. Psychiat.* **41**, 72-80.

Roth, A., and Eddy, E.M. (1967). "Rehabilitation for the Unwanted". Atherton Press, New York.

Ryback, R.S., Lewis, O.F., and Lessard, C.S. (1971a). *Aerosp. Med.* **42**, 529-535.

Ryback, R.S., Trimble, R.W., Lewis, O.F., and Jennings, C.L. (1971b). *Aerosp. Med.* **42**, 408-415.

Schmid, P.G., McCally, M., Piemme, T.E., and Shaver, J.A. (1971). *In* "Hypogravic and Hypo-dynamic Environments", (R.H. Muray and M. McCally, eds.), pp. 211-224. NASA SP-269, National Aeronautics and Space Adminisration, Washington, DC.

Sells, S.B. and Berry, C.A. (1961). "Human Factors in Jet and Space Travel", 330 pp. Ronald Press, New York.

Shulte, J.H. (1967). *Arch. Environ. Health* **14**, 333-336.

Smith, S. (1969). *In* "Sensory Deprivation: Fifteen Years of Research" (J.P. Zubek, ed.), pp. 374-403. Appleton Century-Crofts.

Sorokin, P.A., Simonenko, V.V., and Korolev, B.A. (1969). *In* "Problems of Space Biology," Vol. 13, (A.M. Genin and P.A. Sorokin, eds.). Nauka Press, Moscow. (In English, NASA TT F-639, pp. 12-25).

Spencer, W.A., Vallbona, C., and Carter, R.E. (1965). *Arch. Phys. Med. Rehab.* **46**, 1.

Steinberg, F.U. (1980). "The Immobilized Patient: Functional Pathology and Management", 156 p. Plenum Medical Books, New York.

Stepanova, S.I. (1973). *Kosm. Biol. Med.* **7(1)**, 65-72.

Stepanova, S.I. (1977). *In* "Problems of Space Biology: Current Problems in Space Biorhythmology" (Russian), pp. 311, Isdatel'stvo Nauka, Moscow.

Stevens, P.M., Miller, P.B., Gilbert, C.A., Lynch, T.N., Johnson, R.L., and Lamb, L.E. (1966). *Aerosp. Med.* **37**, 357-367.

Storm, W.F. and Gianetta, C.L. (1974). *Aerosp. Med.* **45**, 431-433.

Strange, R.E. and Klein, W.F. (1973). *In* "Polar Human Biology" (O.G. Edholm and E.K.E. Gunderson, eds.), pp. 410-429, Heinemann Medical Books, London.

Stroebel, C.F. (1969). *Bibl. Primat.* **9**, 91-105.

Sussman, M.B. (1964). "Rehabilitation and Tuberculosis." Western Reserve University, Cleveland, Ohio.

Takahashi, Y., Kipnis, D.M., and Daughsday, W.H. (1968). *J. Clin. Invest.* **47**, 2079-2090.

Taranenko, Yu.N. (1981). *Kosm. Biol. Aviakosm. Med.* **15(1)**, 45-48.

Taub, J.M. and Berger, R.J. (1974). *In* "Chronobiology" (L.E. Sheving, F. Halberg, and J.E. Pauly, eds.), pp. 571-575, Igaku Shoin, Tokyo.

Taylor, H.L., Henschel, A., Brozek, J., and Keys, A. (1949). *J. Appl. Physiol.* **2**, 223-239.

Terelak, J. (1984). *Kosm. Biol. Aviakosm. Med.* **18(2)**, 74-77.

Trimble, R.W. and Lessard, C.S. (1970). "Performance Decrement as a Function of Seven Days of Bedrest." USAF School of Aerospace Medicine, Aerospace Medicine Division Report No. SAM-TR-70-56, pp. 4, Brooks AFB, Texas.

Vallbona, C., Spencer, W.A., Vogt, F.B. and Cardus, D. (1965). "The Effect of Bedrest on Various Parameters of Physiological Function. Part IX: The Effect on the Vital Signs and Circulatory Dynamics". NASA-CR-179. Texas Institute for Rehabilitation and Research, Houston, Texas.

Vanyushina, Yu.V., Gerd, M.A., and Panferova, N.Ye. (1966). *In* "Problems in Aerospace Medicine: Data on the Conference of 24-27 May, 1966" (V.V. Parin, ed.), pp. 103-107. Ministry of Health of the USSR, 1966. (In English, NTIS No. JPRS-33272.)

Vernikos-Danellis, J. and Winget, C.M. (1979). *In* "Chronopharmacology" (A. Reinberg and F. Halberg, eds.), *Adv. Biosci.* **19**, 101-106. Pergamon Press, New York.

Vernikos-Danellis, J., Leach, C.M., Winget, C.M., Rambaut, P.C. and Mack, P.B. (1972). *J. Appl. Physiol.* **33**, 644-648.

Vernikos-Danellis, J., Winget, C.M., Leach, C.S. and Rambaut, P.C. (1974). "Circadian, Endocrine, and Metabolic Effects of Prolonged Bedrest: Two 56-Day Bedrest Studies," NASA TMX-3051, 45 p. National Aeronautics and Space Administration, Washington, DC.

Vernikos-Danellis, J., Leach, C.S., Winget, C.M., Goodwin, A.L. and Rambaut, P.C. (1976). *Aviat. Space Environ. Med.* **47**, 583-587.

Vernikos-Danellis, J., Dallman, M.F., Forsham, P., Goodwin, A.L. and Leach, C.S. (1978). *Aviat. Space Environ. Med.* **49**, 886-889.

Volkov, M.Yu. (1979). *In* "Proceedings, Aerospace Medicine Sixth All-Union Conference on Space Biology and Aerospace Medicine, Kaluga, June 1979", Vol. II (Russian), pp. 194-195.

Walters, R.H. and Henning, G.B. (1961). *Aerosp. Med.* **32**, 431-434.

Webb, W.B. (1968). "Sleep. An Experimental Approach." MacMillen, New York.

Wever, R. (1979). "The Circadian System of Man, Results of Experiments Under Temporal Isolation", 276 p. Springer-Verlag, New York.

Wever, R.A. (1980). *Chronobiology* **7**, 303-327.

Wever, R.A. (1983a). *Pflügers Arch.* **396**, 85-87.

Wever, R.A. (1983b). *Pflügers Arch.* **396**, 128-137.

Wever, R.A. (1984a). *Experientia* **40**, 1226-1234.

Wever, R.A. (1984b). *Sleep* **7**, 27-51.

Wexler, D., Mendelson, J., Leiderman, P.H., and Solomon, P. (1958). *Arch. Neurol. Psychiat.* **79**, 225-233.

Weybrew, B.B. (1957). "Psychological-Psychophysical Effects of Long Periods of Submergence." U.S. Naval Medical Research Laboratory, Report No. 281, Naval Submarine Base, Connecticut.

Weybrew, B.B. (1962). "Prediction of Adjustment to Long Submergence Aboard a Fleet Ballistic Missile Submarine. Interrelationships of the Adjustment Criteria." USN Submarine Medical Research Laboratory Report No. 383.

Weybrew, B.B. (1971). "Submarine Crew Effectiveness During Submerged Missions of Sixty or More Days Duration", pp. 29. Naval Submarine Medical Center Report NSMRL-686, Groton, Connecticut.

Wheaton, J.L. (1959). "USAF SAM Aeromedical Reviews", pp. 5-59.

Williams, B.A. and Reece, R.D. (1972). *Aerosp. Med. Assoc. Preprints*, pp. 140-141.

Winget, C.M., Vernikos-Danellis, J., DeRoshia, C.W., and Cronin, S. (1974b). *In* "Biorhythms and Human Reproduction" (M. Ferin, F. Halberg, R.M. Richart, and R.L. Van de Wiele, eds.), pp. 575-587. Wiley, New York.

Winget, C.M., DeRoshia, C.W., and Beljan, J.R. (1974a). *Aerosp. Med. Assoc. Preprints*, pp. 87-88.

Winget, C.M., Vernikos-Danellis, J., DeRoshia, C.W., and Cronin, S. (1974b). *In* "Biorhythms and Human Reproduction" (M. Ferrin, ed.), pp. 575-587. Wiley, New York.

Winget, C.M., Vernikos-Danellis, J., Leach, C.S., and Rambaut, P.C. (1974c). *In* "Chronobiology" (L.E. Scheving, F. Halberg, and J. Pauly, eds.), pp. 429-434, Igaku Shoin, Tokyo.

Winget, C.M., DeRoshia, C.W., and Sandler, H. (1976). *Aerosp. Med. Assoc. Preprints*, pp. 254-255.

Winget, C.M., DeRoshia, C.W., Vernikos-Danellis, J., Rosenblatt, L.S., and Hetherington, N.W. (1977). *Waking and Sleeping* **1**, 359-363.

Winget, C.M., DeRoshia, C.W., and Sandler, H. (1979). *The Physiologist* **22(Suppl.)**, S79-S80.

Wright, J.E., Vogel, J.A., Sampson, J.B., Knapik, J.J., Patton, J.F., and Daniels, W.L. (1983). *Aviat. Space Environ. Med.* **54**, 132-137.

Yefimenko, G.D. (1969). *In* "Problems of Space Biology", Vol. 13, Prolonged Limitation of Mobility and its Influence on the Human Organism" (A.M. Genin and P.A. Sorokin, eds.), pp. 121-132. Nauka Press, Moscow. (NASA TT F-639.)

Yurchenko, A.A. (1981). *Kosm. Biol. Aviakosm. Med.* **15(2)**, 84-87.

Zav'yalov, Ye.S. and Mel'nik, S.G. (1967). *Kosm. Biol. Med.* **1(3)**, 57-62.

Zav'yalov, Ye.S., Mel'nik, S.G., Chugunov, G.Ya., and Vorona, A.A. (1970). *Kosm. Biol. Med.* **4(1)**, 61-65.

Zimmerman, J., Czeisler, C.A., and Weitzman, E.D. (1981). "Chronobiology of Performance and Moods During Temporal Isolation in Humans." Paper presented at A.P.S.S. Meeting, Hyannis, Massachusetts.

Zubek, J.P. (1968). *J. Abnorm. Psychol.* **73**, 223-225.

Zubek, J.P. (1973). *In* "Man in Isolation and Confinement" (J.E. Rasmussen, ed.), pp. 9-83. Aldine Publishing Co., Chicago, Illinois.

Zubek, J.P. and MacNeill, M. (1966). *Canad. J. Psychol.* **20**, 316-336.

Zubek, J.P. and Wilgosh, L. (1963). *Science* **140**, 306-308.

Zubek, J.P., Bayer, L., Milstein, S., and Shephard, J.M. (1969a). *J. Abnorm. Psychol.* **74**, 230.

Zubek, J.P., Bayer, L., and Shephard, J.M. (1969b). *J. Abnorm. Psychol.* **74**, 625-631.

Zulley, J., Wever, R., and Aschoff, J. (1981). *Pflugers Arch.* **391**, 314-318.

7

Exercise Responses After Inactivity

VICTOR A. CONVERTINO

The Bionetics Corporation
Biomedical Operations and Research Office
Kennedy Space Center, Florida 32899

I. INTRODUCTION

Exercise represents an acute environmental stressor that induces several physiological responses such as increases in heart rate, oxygen consumption, minute ventilation volume, cardiac output, and sweat rate in order to sustain the challenge of muscular work. The unique and significant aspect of studying physiological mechanisms by examining the individual's adaptation to muscular exercise was best described in 1934 by Joseph Barcroft:

> Sometimes as I have stood contemplating the majesty of a locomotive by the platform of a railway station, I have thought how meaningless would be the machine unless considered in respect of its activity. . . . apart from the fulfillment of its function, the engine is an agglomeration of curiously shaped pieces of metal. The condition of exercise is not a mere variant of the condition of rest, it is the essence of the machine.

Based on this premise, exercise represents an environmental stimulus to provide baseline stimulation for normal life function. With this stimulation, the organism maintains a physiological reserve that provides the capacity to meet the demands of walking, mowing the lawn, vacuuming the carpet, and other normal daily activities. Without the physiological stimulation from regular muscular activity, the physiological reserve may be significantly reduced, and the individual's capability to perform normal daily activities will be minimized. Indeed, we observe this contrast between the trained athlete, who performs a level of work with very little strain, and the individual who has been confined to bed and cannot sustain the same work level for any considerable time. Thus, exercise provides an experimental condition in which we can study longitudinal effects of experiment or condition duration. Thus, exercise provides a means for docu-

149

menting the dynamic characteristics of physiological mechanisms under the body's most inherent functional condition—human movement. Furthermore, the evaluation of physical and physiological performance during submaximal and maximal work after prolonged exposure to an environmental change such as bed rest provides a sensitive tool for determining how the adaptation occurs.

The research emphasis in exercise physiology has been aimed at the physiological mechanisms of response to muscular work and the adaptations that occur following chronic repeated exposure, i.e., training. Indeed, if exercise is repeated for weeks or months, the capacity to perform work is increased while the stress on the systems of the body is diminished. Furthermore, these biological adaptations are specific to the level of training, that is, the intensity, duration, and frequency of the exercise. The volume of knowledge acquired through the study of exercise training has contributed significantly to our understanding of the physiological adaptations associated with chronic challenge of increased muscular activity and its many implications for health maintenance, as well as for athletic performance. However, little attention has been given to the opposite end of the activity adaptation continuum—the model of deconditioning induced by chronic inactivity.

Chronic exposure to bed rest provides an interesting and unique experimental model for study, since it induces a state of inactivity and a physiological state that is the exact opposite of that induced by physical conditioning. During bed rest, the body is exposed to several environmental disturbances including (a) a reduction in the hydrostatic pressure gradient within the cardiovascular system; (b) a reduction of weight load upon the muscles and bones; (c) a reduction in total body activity and energy requirement or output, or both; and (d) the problems of isolation and confinement. The determination of physiological responses to exercise following bed rest can be a helpful tool to the exercise physiologist, aerospace scientist, and clinician in understanding the mechanisms of the deconditioning process and in developing countermeasures for health maintenance and physical work capacity under these conditions. Results have direct application to (a) weightlessness exposure, (b) prolonged hospital confinement, (c) sedentary life-styles, and (d) the aging process.

II. EXERCISE PERFORMANCE AFTER BED REST

A. Maximal Oxygen Uptake

It is well documented that a general deconditioning is induced by prolonged bed rest, an adaptive process that affects the functioning of most physiological systems of the body. Some physiological changes associated with chronic bed rest exposure include reductions in total body fluids, electrolytes, plasma volume, red cell volume, and hemoglobin content, as well as decreases in muscle size, tone, and strength. Each of these is known to be able to independently affect measured physical working capacity and maximal oxygen uptake ($\dot{V}_{O_2\ max}$). Therefore, it is not surprising that a reduction in $\dot{V}_{O_2\ max}$ and a subsequent decrease in physical working capacity have been reported from nearly all studies that utilized complete bed or chair rest.

White and associates (1966) reported a mean reduction in $\dot{V}_{O_2\ max}$ during treadmill exercise from 3.66 to 3.47 liter/min (-5.2%) in 3 men who underwent 10 days of continuous bed rest. In comparison, a longer bed rest confinement of 26 days resulted in a decrease in $\dot{V}_{O_2\ max}$ from 2.57 to 2.07 liter/min (-19.5%) in 22 male subjects (Stevens et al., 1966b). Similar results were observed by Meehan and co-workers (1966) in 14 male subjects exposed to 28 days of bed rest, which decreased their $\dot{V}_{O_2\ max}$ from 3.75 to 3.04 liter/min (-18.9%). In another study (Lamb et al., 1964b), 36 men were confined to 14–30 days in a space cabin simulator (chair rest), and $\dot{V}_{O_2\ max}$ decreased from 2.66 to 2.28 liter/min (-14.3%), with a subsequent reduction in physical working capacity, as indicated by a change in maximal treadmill exercise duration from 15.3 to 13.0 min. Kakurin and associates (1966) reported that 20 days of bed rest produced decreases in bicycle ergometer exercise performance in 4 male subjects, as manifested by decreases in maximal work rate (-26.0%), maximal work time (-20.3%), and $\dot{V}_{O_2\ max}$ from 3.10 to 2.70 liter/min (-12.9%). The bed rest study of Saltin et al. (1968) demonstrated the largest reported decrement in the mean treadmill $\dot{V}_{O_2\ max}$ of 26.4% in 5 young men following 20 days of bed rest confinement.

A series of studies from investigations at the National Aeronautics and Space Administration (NASA) Ames Research Center in which supine bicycle ergometry was used to assess intolerance to exercise following bed rest demonstrated reductions in $\dot{V}_{O_2\ max}$ of 7.2, 9.1, and 14.0% following 10, 14, and 15 days of bed rest, respectively (Convertino et al., 1977, 1982b, 1982c). Indeed, in contrast to the increase in $\dot{V}_{O_2\ max}$ and physical working capacity routinely observed with repeated exercise activity, the data presented from bed rest studies corroborate general observations that bed rest inactivity adversely affects the cardiorespiratory response to exercise in normal subjects and hospital-confined patients (Bassey et al., 1972, 1973; Birkhead et al., 1963a,b, 1964a,b; Taylor, 1968; Taylor et al., 1949).

B. Oxygen Uptake during Submaximal Work

Submaximal $\dot{V}_{O_2}$ appears to remain constant following bed rest (Bassey et al., 1973; Convertino et al., 1981, 1982a, 1984; Saltin et al., 1968), suggesting that mechanical efficiency does not change with deconditioning. However, others have reported decreases in submaximal $\dot{V}_{O_2}$, indicating that mechanical efficiency was indeed reduced (Iseyev and Katkovskiy, 1968, Iseyev and Nefedov, 1968; Kakurin *et al.*, 1963). Interestingly, three studies reported that the mean submaximal $\dot{V}_{O_2}$ was consistently lower following prolonged bed rest at all submaximal work levels (Convertino *et al.*, 1977, 1982b; Stremel *et al.*, 1976). This decrease might have resulted from an increased mechanical efficiency, but that would seem unlikely to have occurred after deconditioning. A more plausible explanation involves a change in the time constant (rate change during the transient phase of exercise) for the oxygen uptake to reach an equilibrium level. If the time constant were lengthened after bed rest, then in 2–3 min the oxygen uptake might not have reached a steady-state submaximal value. Thus, a measurement of submaximal $\dot{V}_{O_2}$ could be lower after bed rest without a change in mechanical efficiency.

In recent studies, Convertino and co-workers (1981, 1982a,c, 1984) demonstrated that the mechanical efficiency of bicycle ergometer work was not altered by 7 days of anti-orthostatic bed rest, since the post-bed-rest work output and steady-state energy requirement ($\dot{V}_{O_2}$) were equal to pre-bed rest-values. However, they reported (Convertino and Sandler, 1982; Convertino *et al.*, 1984) a significant reduction in the rate of attaining steady-state $\dot{V}_{O_2}$ at a submaximal work intensity of 65% $\dot{V}_{O_2\ max}$, i.e., a shift of the $\dot{V}_{O_2}$ kinetics curve to the right, as shown in Fig. 1. These results indicated that bed rest results in a reduction in total oxygen transport or utilization, or both, during the transient phase of submaximal exercise. This response is manifested by significant increases both in the oxygen deficit during the initial phase of exercise and in oxygen uptake during recovery, despite the ability to attain similar steady-state $\dot{V}_{O_2}$. It is important to note, however, that the changes in $\dot{V}_{O_2}$ kinetics following bed rest reported by Convertino and co-workers (1984) and Convertino and Sandler (1982) were specific to upright rather than supine exercise, suggesting that this change in $\dot{V}_{O_2}$ during the transient phase of submaximal exercise may be associated with orthostatic factors that could contribute to the limitation of stroke volume maintenance by influencing ventricular filling (Hung *et al.*, 1983).

C. Duration of Bed Rest Inactivity

If the relative (percentage) change in $\dot{V}_{O_2}$ is compared to the duration of bed rest using a compilation of available data from the research literature, as shown in Fig. 2, it appears that a primary factor in determining exercise performance following bed rest is the number of days of confinement. Quantitatively, these data from the literature suggest that we could predict a change in $\dot{V}_{O_2\ max}$ of approximately 0.8% per day of bed rest. This relationship must be approached with caution, however, since a number of interacting factors other than the duration of inactivity could affect the magnitude of change in $\dot{V}_{O_2\ max}$. For instance, it has been demonstrated that the change in $\dot{V}_{O_2\ max}$ and other car-

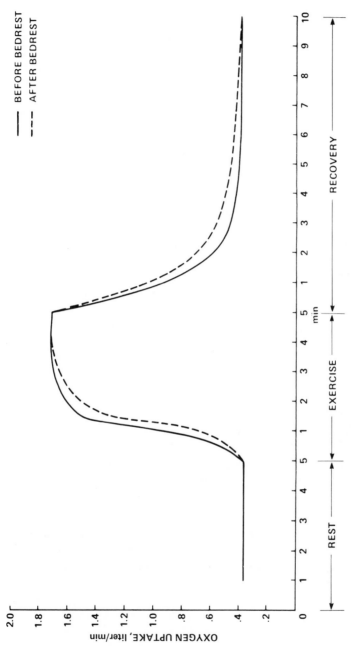

Fig. 1. $\dot{V}_{O_2}$ kinetics before and after bed rest during constant-load exercise of 700 kg-m/min (modified from Convertino *et al.*, 1984a).

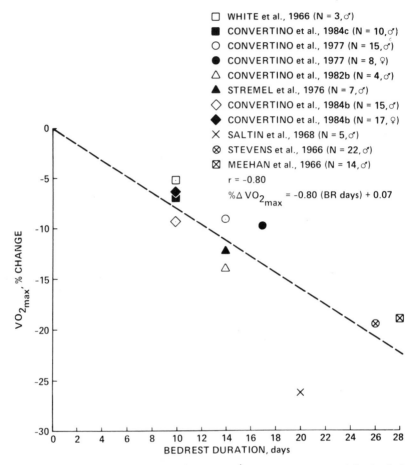

Fig. 2. Relationship between bed rest duration and $\dot{V}_{O_2\,max}$ change (percent) following bed rest (compilation of available data from 11 published studies).

diorespiratory responses following bed rest were significantly different when the same subjects performed the exercise in the upright compared with the supine position (Convertino 1983a,b; Convertino *et al.*, 1981, 1982a,b). Furthermore, in these same subjects, the $\dot{V}_{O_2}$ at volitional fatigue during exercise was significantly lowered when the increments of work rate were increased (Convertino *et al.*, 1982b). Thus, the change in $\dot{V}_{O_2\,max}$ may be due in part to interstudy variations in the type of confinement (chair rest, bed rest, immobilization, etc.), as well as in the exercise mode (bicycle, treadmill, trampoline, static strength, etc.) and test protocol (upright, supine, etc.), used to elicit the $\dot{V}_{O_2\,max}$. Another discrepancy for the relationship between duration of bed rest inactivity and reduction in $\dot{V}_{O_2 \, max}$ results from findings of two studies that demonstrated no significant change in $\dot{V}_{O_2 \, max}$ and physical working capacity following 6–15 days of bed rest (Chase et al., 1966; Lamb et al., 1965).

An additional factor that makes it difficult to evaluate the actual effect of bed rest duration on the change in $\dot{V}_{O_2 \, max}$ and physical working capacity is that most investigators have measured $\dot{V}_{O_2 \, max}$ before and after bed rest. The effect of bed rest duration can be more accurately evaluated if $\dot{V}_{O_2 \, max}$ measurements were performed in different subject groups at various times, e.g., every 2 or 3 days, during the bed rest period. Lamb and co-workers (1965) performed such an experiment using four groups of 6 subjects each; groups 1, 2, 3, and 4 underwent chair rest inactivity for 4, 6, 8, and 10 days, respectively. These investigators found that the reduction in $\dot{V}_{O_2 \, max}$ between chair rest of 4 and 8 days was similar (6.9–7.7%). Interestingly, they found no change in $\dot{V}_{O_2 \, max}$ in the group with 6 days of chair rest, and a 9.7% increase in $\dot{V}_{O_2 \, max}$ in the 10-day group.

In a group of 4 subjects, Georgiyevskiy and associates (1966) reported that $\dot{V}_{O_2 \, max}$ decreased from a pre-bed-rest value of 3.14 liter/min to 2.87 liters per min by 2–3 days of bed rest and remained at 2.87 and 2.77 liter/min at days 13 and 20, respectively. These data suggest that in contrast to the hypothesis that the duration of bed rest is a primary factor in $\dot{V}_{O_2 \, max}$ decrements the largest proportion of the change in $\dot{V}_{O_2 \, max}$ and exercise performance that accompanies bed rest deconditioning may occur within the first hours to days of confinement and therefore may be related to general and rapid physiological adaptation of systems. This relationship is further supported by the observations that 6 hr of water immersion, a method used to simulate the acute exposure to weightlessness, produced similar decreases in $\dot{V}_{O_2 \, max}$ (12–18%) to those observed following 14–28 days of bed rest (Stegemann et al., 1969). Therefore, although a relationship exists between the duration of bed rest inactivity and the relative reduction in $\dot{V}_{O_2 \, max}$ and exercise performance, as indicated by cross-sectional studies, there are data to suggest that under a specific bed rest condition, the decrease in $\dot{V}_{O_2 \, max}$ is best correlated with fluid/volume changes that occur within the first hours to days of confinement.

D. Age and Gender

The early investigations of bed rest used young healthy male subjects, ages 18–26 years. Although this approach provided useful information concerning the general effects of bed rest inactivity on exercise performance, the question remained whether the $\dot{V}_{O_2 \, max}$ and work capacity of older subjects and females deteriorated to the same extent. With regard to gender, this problem has important practical implications, since women are increasing their participation in athletics, physical exercise, and the work force, as well as taking part in spaceflight. It is equally important to obtain data on men and women in middle age, since sedentary living habits, as well as clinical problems of hospital confinement, are most common in these populations.

Two studies have specifically investigated the differences between physiological responses to exercise before and after bed rest inactivity in men and women. Convertino and co-workers (1977) compared the changes in $\dot{V}_{O_2 \ max}$ and physical working capacity of young men (19–23 years) and women (23–34 years) following 14 and 17 days of continuous bed rest, respectively. In a subsequent series of experiments (Convertino et al., 1986a), $\dot{V}_{O_2 \ max}$ and physical working capacity of middle-aged men and women (45–65 years) were determined before and after 10 days of bed rest. The comparisons between these four groups are presented in Table I. It is well established that $\dot{V}_{O_2 \ max}$ and physical working capacity are greater in men than in women and decrease with age (Astrand, 1960). Despite this significant difference in pre-bed-rest $\dot{V}_{O_2 \ max}$ and physical working capacity between the men and women and the young and middle-aged subjects, the relative (percentage) changes in $\dot{V}_{O_2 \ max}$ and physical working capacity resulting from bed rest inactivity were very similar in each group. The $\dot{V}_{O_2 \ max}$, measured by supine bicycle ergometry, decreased by 9.1% in the young men; 9.7% in the young women; 9.4% in the middle-aged men; and 7.3% in the middle-aged women. These changes in $\dot{V}_{O_2 \ max}$ are similar to those reported from previous studies in young males (Georgiyevskiy et al., 1966; Lamb et al., 1965; White et al., 1966) and middle-aged men (Convertino et al., 1982b). The corresponding changes in maximal work rate and maximal exercise duration, both indices of functional working capacity, were not significantly different between any of the groups (Table I). Thus, it is clear that despite the significant differences in the absolute working capacity between men and women and subjects of varying ages, the proportional deconditioning, i.e., adaptation, to bed rest inactivity is essentially the same for most individuals.

Bed rest deconditioning does not change submaximal $\dot{V}_{O_2}$ in young men (Bassey et al., 1973; Convertino et al., 1981, 1982a,c, 1984; Convertino and Sandler, 1982; Saltin et al., 1968). In contrast, the young women demonstrated a decrease in $\dot{V}_{O_2}$ at submaximal work rates that suggests they have an additional factor present that restricts oxygen utilization. Since maximal heart rates between men and women were essentially the same, the women appear to have no selective limitation of heart function (a central effect). One plausible explanation involves a change in the kinetics of oxygen uptake, i.e., the time constant for the $\dot{V}_{O_2}$ to reach an equilibrium level. If the time constant were lengthened after bed rest, then the $\dot{V}_{O_2}$ might not reach a steady state by the time it is measured and, as a result, the submaximal $\dot{V}_{O_2}$ following bed rest would be lower. Indeed, Convertino and co-workers (1984) demonstrated that the $\dot{V}_{O_2}$ kinetics curve shifts to the right following bed rest to lengthen the time constant. This kinetic change might be greater in women than in men. However, similar percentages of change in $\dot{V}_{O_2, \ max}$ between men and women would suggest that these submaximal

TABLE I

Mean (±SE) Values and Changes in $\dot{V}_{O_2max}$ and Physical Work Capacity in Young and Middle-Aged Men and Women before and after Bed Rest[a]

Subjects	No. of subjects	Age (yr)	$\dot{V}_{O_2 max}$ STPD (liter/min)[b]			Maximal work rate (watts)[b]			Test duration (min)[b]		
			Pre Br	Post BR	% Δ	Pre BR	Post BR	% Δ	Pre BR	Post BR	% Δ
Young men	15	21 ±1	3.52 ±.12	3.20 ±.12	-9.1*	303 ±13	297 ±11	-2.0	13.0 ±.6	12.1 ±.5	-6.9*
Young women	8	28 ±2	2.06 ±.13	1.86 ±.12	-9.7*	145 ±7	140 ±8	-3.4	12.7 ±.6	12.2 ±.6	-3.2*
Middle-aged men	15	55 ±2	2.78 ±.16	2.52 ±.11	-9.4*	168 ±11	164 ±9	-2.4	12.4 ±.8	11.4 ±.9	-8.1*
Middle-aged women	17	55 ±1	1.65 ±.07	1.53 ±.07	-7.3*	108 ±4	106 ±5	-1.9	9.6 ±.4	8.9 ±.5	-7.3*

[a]Modified from Convertino et al. (1984b).

[b]Symbols and abbreviations: Pre BR, before bed rest; Post BR, after bed rest; % Δ, percentage of change; values followed by an asterisk, $p < .05$.

changes have apparently little functional significance on the maximal working capacity.

Another important factor to consider in studying the responses of women to exercise following bed rest is the interaction with the female menstrual cycle. In a bed rest study of young women (Convertino *et al.*, 1977) all but 2 of the women had their menstrual periods before or after the bed rest periods, but there appeared to be no consistent effect of menstruation on the $\dot{V}_{O_2 \text{ max}}$ response. The middle-aged women bed rested later were all postmenopausal. Thus, the effect of menses on exercise performance following bed rest has not yet been determined and requires further investigation.

E. Level of Aerobic Fitness

The magnitude of increase in $\dot{V}_{O_2 \text{ max}}$ during endurance exercise training is inversely related to the initial level of aerobic fitness ($\dot{V}_{O_2 \text{ max}}$) before training began—that is, the more fit an individual is when he or she commences an exercise training program, the less $\dot{V}_{O_2 \text{ max}}$ will increase as a result of the training (Saltin *et al.*, 1968). From the perspective of adaptation, these data suggest that the individual with a high level of fitness is near his or her maximal physiological reserve. Based on this relationship, it has been proposed that a subject with high $\dot{V}_{O_2 \text{ max}}$ will exhibit a greater percentage of loss of maximal aerobic capacity following bed rest than an individual with an initially low $\dot{V}_{O_2 \text{ max}}$ (Saltin *et al.*, 1968). This relationship has far-reaching implications for experimental or clinical bed rest, sedentary life-styles, aging, and spaceflight. For example, the astronauts who undertake regular aerobic training programs will be predisposed to a greater loss of working capacity during and following spaceflight. Furthermore, it might be predicted that fit individuals who become sedentary will reduce their physical working capacity to a much greater extent than individuals who already live sedentary life-styles.

The earliest data that were consistent with the preceding hypothesis came from 2 healthy men who underwent 28 days of bed rest (Taylor *et al.*, 1949). The change in $\dot{V}_{O_2 \text{ max}}$ measured on the treadmill for the higher capacity subject (initial $\dot{V}_{O_2 \text{ max}} = 4.15$ liters per min) was -22% compared with -13% observed in the lower-capacity individual man (initial $\dot{V}_{O_2 \text{ max}} = 3.54$ liters per min). However, a review by Greenleaf and Kozlowski (1982) pointed out that data from the literature do not clearly support this hypothesis. Whereas one study reported a correlation of -0.78 between pre-bed-rest $\dot{V}_{O_2 \text{ max}}$ and the percentage of change in $\dot{V}_{O_2 \text{ max}}$ after bed rest (Convertino *et al.*, 1977), other studies found poor correlations between these variables (Chase *et al.*, 1966; Saltin *et al.*, 1968). However, a number of factors must be considered in evaluating these data. The studies that produced the poorest correlations (Chase *et al.*, 1966; Saltin et al., 1968) observed data from only 4 and 5 subjects, whereas the study of Convertino and associates (1977) reported results on 15 subjects. A still more important factor may be body position during the $\dot{V}_{O_2 \ max}$ exercise test. The smallest variations in results occur with supine bicycle exercise, in which orthostatic factors are minimal, as opposed to upright bicycle ergometry or treadmill exercise (Greenleaf and Kozlowski, 1982).

In a series of studies performed at the NASA Ames Research Center from 1971 through 1978, 15 young (19–23 years) and 15 middle-aged (45–65 years) men performed similar graded $\dot{V}_{O_2 \ max}$ tests using supine bicycle exercise before and after 10–14 days of bed rest (Convertino et al., 1984a). In both groups, there was a significant negative correlation between the initial pre-bed-rest $\dot{V}_{O_2 \ max}$ and the percentage of change in $\dot{V}_{O_2 \ max}$, suggesting that habitual levels of physical activity and, thus, levels of physical conditioning appear to influence the degree of deconditioning seen after bed rest (see Fig. 3). These findings have been corroborated by a more recent study of 10 middle-aged subjects following 10 days of bed rest (Convertino et al., 1986b). The pre-bed-rest values of $\dot{V}_{O_2 \ max}$ were substantially larger in the young men (47.2 ml/kg/min) than those of the

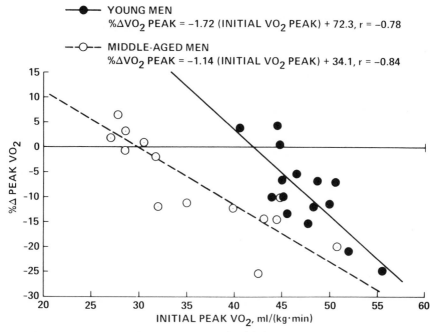

Fig. 3. Relationship between initial pre-bed rest $\dot{V}_{O_2 \ max}$ (peak) and the percent change in $\dot{V}_{O_2 \ max}$ following bed rest in young and middle-aged male subjects (modified from Convertino et al., 1984b).

middle-aged men (35.6 ml/kg/min), but the percentages of reduction in $\dot{V}_{O_2\ max}$ were similar (9.3 and 8.4%, respectively), suggesting that the relative rate of bed rest deconditioning is similar in young and middle-aged subjects and depends more on initial $\dot{V}_{O_2\ max}$ than on age. The slope of the regression line between initial $\dot{V}_{O_2\ max}$ and post-bed-rest change in $\dot{V}_{O_2\ max}$ was significantly steeper in the young subjects than in the older men (Fig. 3), suggesting that the rate of absolute reduction in work capacity during bed rest in the absence of regular physical activity is greater in younger subjects. It is interesting to note that this relationship between initial $\dot{V}_{O_2\ max}$ and change in $\dot{V}_{O_2\ max}$ following bed rest is not nearly as strong in female subjects (Convertino et al., 1984a).

Therefore, the data suggest that an individual with an initially high work capacity and peak $\dot{V}_{O_2\ max}$ experiences a larger reduction in peak $\dot{V}_{O_2}$ following bed rest inactivity. The less marked reduction in peak $\dot{V}_{O_2}$ in absolute terms for older subjects must be partly explained by an effect of aging; however, the middle-aged person appears to have the same relative ability as younger subjects to reduce his or her aerobic power following prolonged bed rest.

F. Muscle Strength

A compilation of available data on changes in strength of the major muscle groups induced by bed rest deconditioning is presented in Table II, as modified from the work of Greenleaf and associates (1983). All subjects were men, with bed rest periods ranging from 7 to 120 days. Mean reductions in handgrip, forearm, and arm strengths (−5% to −11%) were about one-half as great as the strength losses in the trunk and leg muscle groups (−12% to −24%). Although Panov and Lobzin (1968) reported a progressive decrease in handgrip strength after 64 days of bed rest without remedial exercise, analysis of all data suggests that strength in both small-muscle (arm, forearm, hand) and large-muscle (trunk, thigh, leg) groups does not substantially decrease until after 2 weeks of bed rest. Although reductions in neuromuscular performance may not occur within the first 7 days of bed rest (Hargens et al., 1983; Trimble and Lessard, 1970), regression of changes in maximal muscular strength over the duration of 72 days of bed rest suggests loss of strength in all muscle groups at a rate of 0.7% per day. This is particularly the case for bed rest periods longer than 2 weeks when remedial exercise procedures are not employed (Greenleaf et al., 1983). The reduction in muscular strength following bed rest is associated with decreased muscle electrical activity and increased fatigability (LaFevers et al., 1977). Inactivity is the apparent prime cause for these changes, since bicycle ergometer isotonic leg exercise training during bed rest resulted in essentially no detectable loss of strength in all measured muscle groups (Kakurin et al., 1970; Yeremin et al., 1969).

III. PHYSIOLOGIC CHANGES ASSOCIATED WITH REDUCED WORK CAPACITY

A. Pulmonary Function

Reduction in pulmonary function with a consequent impedance of maximal gas exchange might be one mechanism by which $\dot{V}_{O_2\ max}$ and thus physical working capacity could be diminished following bed rest deconditioning. Saltin and co-workers (1968) reported that bed rest deconditioning and reconditioning by exercise training did not change total lung capacity, forced vital capacity, or 1-sec forced expiratory capacity and that diffusing capacity of the lungs during exercise showed a tendency to decrease following bed rest. Furthermore, the residual lung volume was not altered by 14 days of bed rest inactivity (Greenleaf et al., 1977). Mean minute ventilation volume has been reported to decrease following bed rest, but this response has been directly related to the decrease in $\dot{V}_{O_2\ max}$ and physical working capacity (Convertino et al., 1982c; Saltin et al., 1968). However, most studies have shown that maximal minute ventilation volume for men and women did not change or increase following bed rest, despite decreases in $\dot{V}_{O_2\ max}$ (Convertino et al., 1977, 1982b, 1986a; Stremel et al., 1976). Furthermore, minute ventilation volume was significantly elevated during equal submaximal work rates and $\dot{V}_{O_2}$ following bed rest (Convertino et al., 1981, 1982a). Despite the close relationship of ventilation to $\dot{V}_{O_2}$ and heart rate during submaximal and maximal work in ambulatory subjects, the minute ventilation volume appears to be independent of the deconditioning effect of bed rest. Thus, there is little evidence to suggest that changes in pulmonary function can account for the commonly observed reduction in $\dot{V}_{O_2\ max}$ and work capacity induced by bed rest deconditioning.

B. Blood Volume

Changes in physiological responses to exercise following bed rest have been associated with changes in vascular fluid volume. The $\dot{V}_{O_2\ max}$ and physical working capacity following deconditioning appear to be proportionately affected by decreased plasma and total blood volume (hypovolemia). In two separate studies, reductions in $\dot{V}_{O_2\ max}$ of 7.0 and 14.0% were associated with mean plasma volume losses from 3627 to 3229 ml (11.0%) in 10 subjects after 10 days of bed rest (Convertino et al., 1986b) and from 4060 to 3380 ml (16.7%) in 4 individuals who underwent 14 days of bed rest (Convertino et al., 1982c). Convertino and associates (1977) reported similar plasma volume reductions of 11.3% in 15 men and 12.6% in 8 women following similar bed rest exposures, with decreases in $\dot{V}_{O_2\ max}$ of 9.1 and 9.7% in men and women, respectively. Following 14 days of continuous bed rest, 7 young male subjects had plasma

TABLE II

Mean Percentage of Change in Maximal Strength of Various Muscle Groups after Bed Rest with and without Supine Exercise Training during Bed Rest[a]

Reference	No. of male subjects	Bed rest exercise schedule			Muscle group strength	
		No. of days bed rest	Exercise duration (min/day)	Exercise mode[b]	Muscle group	% Δ
Friman and Hamrin (1976)	14	7	None	None	Handgrip	-5
					Thigh	-4
					Posterior leg	-7
Trimble and Lessard (1970)	8	7	None	None	Handgrip	0
Greenleaf et al. (1983)	7	14	None	None	Handgrip	—
			60	ITE[c]	Handgrip	-1
			60	IME	Handgrip	+1
Taylor et al. (1949)	6	21	None	None	Handgrip	-3
					Back	-8
Birkhead et al. (1964b)	2	24	60	ITE[c]	Arm	-2
Birkhead et al. (1963b)	4	42	60	ITE[c]	Arm	-5
Deitrick et al. (1948)	4	42–49	None	None	Handgrip	0
					Forearm	-7
					Arm	-9
					Back	0
					Abdomen	0
					Anterior leg	-13
					Posterior leg	-21
Kakurin et al. (1970)	3	62	None	None	Back	-19
	3	62	150	ITE[c]	Back	-9

Study						
Yeremin et al. (1969)	1	70	None	None	Forearm	−26
					Arm	−28
					Back	−39
					Abdomen	−48
					Thigh	−36
					Anterior leg	−57
					Posterior leg	−37
	3	70	120	ITE[c]	Forearm	+1
					Arm	+1
					Thigh	−6
					Anterior leg	−11
					Posterior leg	+1
	3	70	120	ITE[d]	Forearm	+4
					Arm	+1
					Thigh	−3
					Anterior leg	−6
					Posterior leg	+3
Panov and Lobzin (1968)	4	11	None	None	Handgrip	0
		22				−2
		36				−8
		44				−12
		64				−27
Krupina and Tizul (1971)	10	120	None	None	Forearm	−26

[a]Compilation of data from Greenleaf et al. (1983).
[b]ITE, isotonic exercise; IME, isometric exercise.
[c]Isotonic exercise with a cycle.
[d]Isotonic exercise on a treadmill.

volume losses from 3491 to 2966 ml (15.1%), with $\dot{V}_{O_2\ max}$ decreases of 12.3% (Stremel et al., 1976). White and associates (1966) reported a decrease of 5.2% in $\dot{V}_{O_2\ max}$ and reductions in plasma volume from 3285 to 2796 ml (14.9%) in 3 subjects following 10 days of absolute bed rest. In comparison, in longer-duration bed rest studies of 26 days (Stevens et al., 1966b) and 28 days (Meehan et al., 1966), greater decreases in plasma volume from 2911 to 2292 ml (21.3%) and from 3366 to 2432 ml (27.7%) were associated with $\dot{V}_{O_2\ max}$ reductions of 19.5 and 18.9%, respectively. When the percentage of change in plasma volume following bed rest was compared with the percentage of change in $\dot{V}_{O_2\ max}$, a high correlation existed, suggesting that the loss in vascular fluid volume induced by bed rest may be a primary mechanism for the reduction in the body's capacity to transport or to utilize oxygen for energy production during exercise (see Fig. 4).

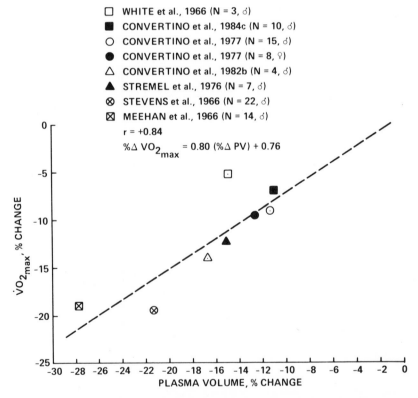

Fig. 4. Relationship between percent change in plasma volume following bed rest and the percent change in $\dot{V}_{O_2\ max}$ (compilation of available data from 8 published studies).

One possible explanation for the effect of vascular hypovolemia on the decrease in $\dot{V}_{O_2\ max}$ after bed rest may be a lower oxygen-carrying capacity of the blood. Indeed, some studies have demonstrated that red cell volume can decrease from 5 to 25% during prolonged bed rest (Convertino *et al.*, 1982a,c; Lamb *et al.*, 1964b; Meehan *et al.*, 1966; White *et al.*, 1966). However, the correlation between the change in red cell volume and the change in $\dot{V}_{O_2\ max}$ is low, since several investigators have reported essentially no change in red cell volume following short-duration bed rest exposure, despite a significant decrease in the $\dot{V}_{O_2\ max}$ (Convertino *et al.*, 1985, 1986b; Stevens *et al.*, 1966b). Furthermore, despite some discrepancies between red cell volume changes in bed rest studies of varying lengths, all studies have demonstrated that hematocrit remains constant or increases slightly, suggesting that the oxygen-carrying capacity per unit volume of blood is not significantly affected by the bed-rest-induced hypovolemia. Therefore, it is unlikely that the primary effect of hypovolemia on reducing the $\dot{V}_{O_2\ max}$ following bed rest results from any significant change in oxygen-carrying capacity per se.

The primary mechanism of the bed-rest-induced reduction in $\dot{V}_{O_2\ max}$ resulting from hypovolemia appears to be the effect of vascular volume change on cardiovascular hemodynamics.

C. Cardiac Function

The impaired $\dot{V}_{O_2\ max}$ and physical working capacity associated with prolonged bed rest involve changes in central cardiac function during exercise, as manifested by elevated heart rate (Deitrick *et al.*, 1948), reductions in stroke volume (Buderer *et al.*, 1976a,b; Olree *et al.*, 1973; Saltin *et al.*, 1968), and decreased ejection time (Buderer *et al.*, 1976a,b). Saltin and co-workers (1968) reported that a 26.4% decrease in the treadmill $\dot{V}_{O_2\ max}$ in 5 young men following 20 days of bed rest resulted from similar reductions in maximal cardiac output from 20.0 to 14.8 liter/min (26.0%) and maximal stroke volume from 104 to 74 ml (28.8%), with essentially no change in maximal heart rate or arteriovenous oxygen difference. Furthermore, stroke volume during supine submaximal exercise (600 kg-m/min) fell 23% compared with a 35% decrease during upright submaximal treadmill exercise of equivalent metabolic demand ($\dot{V}_{O_2}$ of 1.8 liter/min), suggesting that poor postural adaptation, impaired venous return or hypovolemia, or both, could not completely account for the deterioration in exercise capacity following bed rest. The reduction in stroke volume in both supine and upright positions has been corroborated by Katkovskiy and Pometov (1976). Based on these data, it was concluded that the cardiovascular response to muscular exercise following bed rest deconditioning was impaired by a nonspecific deterioration in myocardial function rather than an orthostatic effect (Blomqvist *et al.*, 1971). These investigators hypothesized that the impaired ventricular function may result from cardiac muscle atrophy, as supported by an 11% decline in fluoroscopically measured resting heart volume after bed rest. However, direct comparison of the cardiovascular responses to upright and supine exercise in the study by Saltin *et al.*, (1968) was not possible, since the supine test was submaximal and performed on a bicycle ergometer, whereas the upright test was maximal and performed on a treadmill.

In contrast to the findings of Saltin and associates (1968), Georgiyevskiy *et al.* (1966) found no change in maximal cardiac output following 20 days of absolute bed rest in 4 young men, despite a significant decrease in $\dot{V}_{O_2 \ max}$ and total work performed. In this study, a 6% decrease in stroke volume was compensated for by a 5% increase in maximal heart rate. Similarly, Olree and co-workers (1973) reported a relatively constant cardiac output at constant work rates, despite a decrease in post-bed-rest exercise stroke volume. These results implied that the lower $\dot{V}_{O_2 \ max}$ following bed rest deconditioning did not result from reduced cardiac function but instead from some aspect of oxygen transport or utilization by the tissues, or both.

In more recent studies, the cardiovascular responses to supine and upright posture during submaximal and maximal exercise following bed rest were compared using the same test protocols and work rates on a bicycle ergometer in an attempt to better define the effect of cardiac function on the bed-rest-induced impairment of $\dot{V}_{O_2 \ max}$ and exercise capacity. With slight increases in submaximal and maximal heart rates, middle-aged subjects maintained their $\dot{V}_{O_2}$ during supine exercise at values similar to those measured before bed rest (Convertino *et al.*, 1982b,c). In contrast, these same subjects could not sustain $\dot{V}_{O_2}$ during submaximal and maximal upright exercise, despite even greater increases in exercise heart rates, diastolic pressures, and rate–pressure products (see Fig. 5). In these studies, during a 3-stage graded exercise test consisting of work rates of 250 (stage I), 525 (stage II), and 835 (stage III) kg-m/min, equilibrium-gated cardiac blood pool scintigraphy was performed during supine and upright bicycle ergometry in 10 subjects before and after 10 days of continuous bed rest. The radionuclide data were used to calculate left ventricular ejection fraction and the percentages of relative change in stroke volume from rest to exercise (Hung *et al.*, 1983). In these experiments, there was no evidence of an adverse effect of bed rest inactivity on left ventricular function: left ventricular ejection fraction was augmented during supine and upright exercise following bed rest (Hung *et al.*, 1983; Katkovskiy and Pometov, 1971) (see Fig. 6). The augmentation of both left ventricular ejection fraction and exercise heart rate appeared to be compensatory mechanisms to maintain cardiac output, despite a lower cardiac end-diastolic volume (preload) after bed rest, particularly in the supine posture, since $\dot{V}_{O_2}$ did not change and the arteriovenous oxygen difference may not have changed (Katkovskiy and Pometov, 1971; Saltin *et al.*, 1968). However, during upright exercise, stroke volume showed a tendency to decrease, despite an aug-

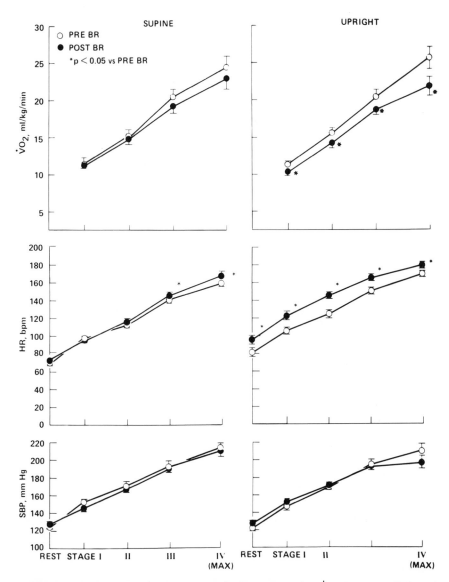

Fig. 5. Responses of submaximal and maximal oxygen uptake ($\dot{V}_{O_2}$), heart rate (HR), and systolic blood pressure (SBP). Mean work rates were equal to 250 (Stage I), 525 (Stage II), 835 (Stage III), and 1165 (Stage IV–max) kg-m/min. Values are mean ± SE (from Convertino *et al.*, 1982a, by permission of the American Heart Association).

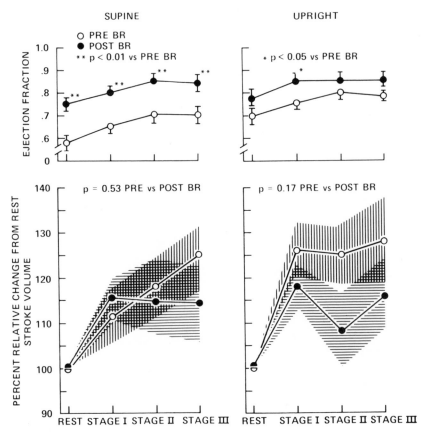

Fig. 6. Changes induced by bed rest in left ventricular ejection fraction and stroke volume during rest and Stages I (250 kg-m/min), II (525 kg-m/min), and III (835 kg-m/min) of exercise. Values expressed are mean ± SE (from Hung *et al.*, 1983, by permission of the American Journal of Cardiology).

mented left ventricular ejection fraction, suggesting that a reduction in exercise cardiac output following bed rest resulted from a lower end-diastolic volume compared with the supine exercise and was not a result of reduced myocardial function.

This major discrepancy between the findings of Hung and co-workers (1983) and Saltin and co-workers (1968) may be explained by several factors. Differences in age and physical conditioning may well explain why supine $\dot{V}_{O_2}$ and cardiac function were better maintained after bed rest in the middle-aged subjects than in the younger ones. Younger, physically fit persons have a higher resting stroke volume, exercise cardiac output, and $\dot{V}_{O_2 \text{ max}}$ than older, relatively sedentary persons (Rowell, 1974). In a later study by Saltin and Rowell (1980) 2 subjects who were best conditioned had the highest values of heart volume and $\dot{V}_{O_2 \text{ max}}$ before bed rest but the greatest absolute decrease in heart volume and $\dot{V}_{O_2 \text{ max}}$ after bed rest. Since the middle-aged subjects in the study of Hung and associates (1983) had initial $\dot{V}_{O_2 \text{ max}}$ values of only 2.5 liter/min, it is expected that they would show smaller decreases in $\dot{V}_{O_2}$, stroke volume, and cardiac output during submaximal and maximal exercise after bed rest than younger or better pre-bed-rest-conditioned subjects, whose initial values of $\dot{V}_{O_2 \text{ max}}$ were 3–4 liter/min. Moreover, the bed rest duration in the study by Saltin and co-workers was twice that of the Hung study, 20 versus 10 days. If myocardial atrophy does occur, it might be that a longer period of time is required for the deteriorating effects on ventricular function to become functionally significant. This hypothesis is consistent with other findings related to loss of skeletal muscle mass and muscle strength and decreased basal metabolic rate (Greenleaf et al., 1983). Although reduction in skeletal muscle protein synthesis leading to atrophy begins within hours of immobilization (Booth, 1977; Booth and Seider, 1979), muscle strength may not exhibit significant reductions until 14–20 days of bed rest exposure.

The mechanisms involved in changes in cardiac function before and after bed rest are not completely clear. Venous pooling after the body assumes the upright posture reduces left ventricular volume and filling pressure (Bevegard et al., 1960; Holmgren and Ovenfors, 1960; Poliner et al., 1980), since an additional 300–800 ml of blood can be sequestered in the legs, compared to what is accumulated in the supine posture (Bevegard et al., 1960). Thus, upright exercise in normal individuals is much more dependent than supine exercise on venous return from the legs and on the Frank-Starling mechanism to augment stroke volume (Bevegard et al., 1960; Poliner et al., 1980). Subsequently, mechanisms for the control of heart rate and stroke volume would become specially sensitive to venous pooling and to underfilling of the heart during upright exercise following bed rest. Further investigations with more precise measurements of end-diastolic volume and stroke volume during supine and upright exercise following bed rest are necessary to quantitatively document the time course of cardiac changes as they affect responses to muscular exercise.

The role of decreased ventricular filling as a primary mechanism in the reduction in exercise cardiac output following bed rest can be further complicated by bed-rest-induced plasma and blood volume reductions, since hypovolemia decreases mean venous pressure, venous return, and stroke volume (Robinson et al., 1966). The importance of this hypovolemic factor in the reduction of cardiac output and $\dot{V}_{O_2 \text{ max}}$ following prolonged bed rest is underscored by a study of Convertino and associates (1982c). A significant decrease in $\dot{V}_{O_2 \text{ max}}$ and an increase in maximal heart rate during upright bicycle ergometry were shown in 4 subjects after 14 days of complete bed rest, while in comparison there was essentially no change in $\dot{V}_{O_2 \text{ max}}$ or exercise heart rate in 4 other subjects who were exposed to 3 hr of daily induced venous pooling throughout bed rest by wearing a reverse-gradient garment designed to simulate the effects of standing (Annis, 1974). Exposure to the simulation of standing enhanced plasma volume retention, which may have contributed to the observed reversal of $\dot{V}_{O_2\ max}$ reduction by sustaining venous return and stroke volume. Such a possibility is further supported by the strong correlation between the percentage of reduction in plasma volume and the percentages of change in maximal heart rate and $\dot{V}_{O_2\ max}$ as shown in Fig. 7.

Based on the data available from the literature, it may be concluded that the significant increase in submaximal and maximal exercise heart rate following prolonged bed rest appears to be the major compensatory mechanism by which cardiac output is maintained despite reductions in venous return and stroke volume. Indeed, plasma volume level is related to maximal heart rate following bed rest (Convertino et al., 1982c) and exercise training (Convertino, 1983b). Elevation of heart rate appears to be the best explanation for the maintenance of exercise cardiac function despite the presence of bed-rest-induced and postural hypovolemia. This is shown in Fig. 8, in which the percentage of change in heart rate as measured in 5 subjects during supine and upright exercise following bed rest can be shown to strongly correlate with the percentage of change in oxygen pulse (the oxygen exchange per heart beat, which is a noninvasive index of stroke volume). However, this compensation is not without cost. The increase in post-bed-rest exercise heart rate produces a significant increase in the rate–pressure product (Convertino et al., 1981, 1982a,b), suggesting a higher myocardial oxygen demand at the same intensity of exercise. This may also prove to be a limiting factor for adequate cardiac function during exercise following prolonged bed rest inactivity.

D. Muscle Tone and Venous Compliance

It has been suggested that changes in muscle or tissue tone in the lower extremities may be an important mechanism associated with bed-rest-induced exercise intolerance, since complete repletion of plasma volume (induced by lower-body negative pressure (LBNP) during bed rest) did not prevent a post-bed-rest decrease in $\dot{V}_{O_2\ max}$ (Stevens et al., 1966a). Indeed, muscle atrophy and reductions in muscle tone following bed rest have been regularly observed and associated with loss of muscle strength (Cherepakhin, 1968; Kakurin et al., 1966, 1970). These changes in muscle tone may result from structural changes in the muscle (i.e., decreased protein synthesis) or changes in neuromuscular responses such as reduced electrical activity (Booth, 1982). Any increase in venous compliance following bed rest would be expected to cause decreases in stroke volume, cardiac output, and arterial pressure due to decreases in venous return (Starling effect) and to become more exaggerated during passive orthostasis and

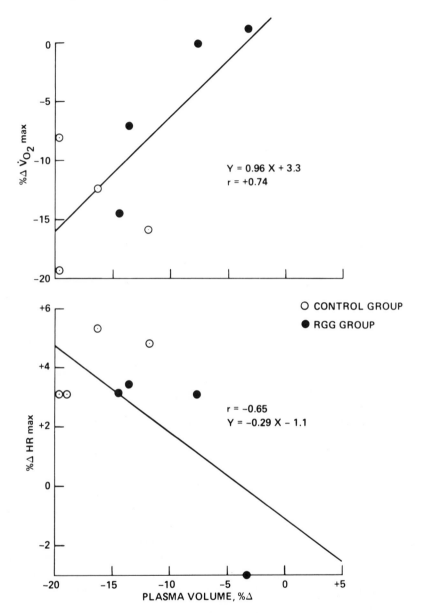

Fig. 7. Relationship between percentage of changes in plasma volume and percentage of changes in maximal oxygen uptake ($\dot{V}_{O_2\ max}$) and maximal heart rate (HR_{max}) (from Convertino *et al.*, 1982b, by permission of the American Physiological Society).

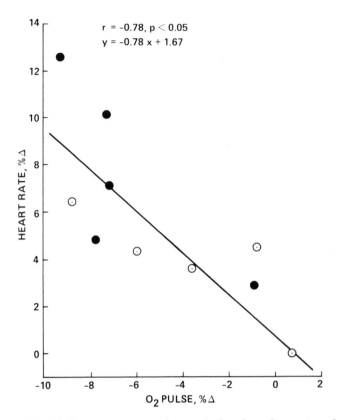

Fig. 8. Relationship between percentage of change in O_2 pulse and percentage of change in heart rate during supine (open circles) and upright (closed circles) exercise following bed rest (from Convertino *et al.*, 1982c).

exercise (Buyanov *et al.*, 1967). This was confirmed by Cooper and Ord (1968) who showed greater decrease in $\dot{V}_{O_2 \, max}$ in subjects during exercise with simultaneous exposure to LBNP, who pooled blood in the lower extremities, as compared to those with no LBNP. The decrease in muscle tone or increase in venous compliance following bed rest deconditioning could compound the effects of hypovolemia during exercise through a reduction in muscular work output or an impairment of venous return, stroke volume, and cardiac output, the result being a reduction in $\dot{V}_{O_2 \, max}$.

E. Cellular Metabolism

A significant adaptation to endurance exercise training is the increased oxidative capacity of muscle fibers as a result of elevated oxidative enzyme activities

(Holloszy and Booth, 1976). Based on this observation, it might be hypothesized that following the prolonged muscular inactivity associated with bed rest, cellular oxidative capacity might be reduced, thus limiting the $\dot{V}_{O_2\ max}$, total energy production rate, and physical working capacity.

The contribution of cellular metabolism as a limiting factor to change in working capacity following bed rest can be examined from the realization that $\dot{V}_{O_2\ max}$ at the systemic level is the product of cardiac output and arteriovenous oxygen difference. Saltin and associates (1968) found that the reduction in $\dot{V}_{O_2\ max}$ following bed rest was explained by a reduction in maximal cardiac output and that there was no change in arteriovenous oxygen difference. These investigators concluded that the reduction in $\dot{V}_{O_2\ max}$ and exercise performance following bed rest resulted from central (cardiac) effects and that there was no alteration in the oxidative capacity of the tissue. In contrast to these results, Georgiyevskiy and co-workers (1966) found no change in maximal cardiac output with a significant decrease in $\dot{V}_{O_2\ max}$, suggesting that after bed rest, the reduction in work capacity resulted from a change in oxygen transport or utilization by the tissues, or both. This may be caused partly by a reduction in capillary density in the muscle (Saltin and Rowell, 1980) or by decreased enzyme capacity. The differences in these two studies remains unsolved.

During graded exercise, the threshold of blood lactate accumulation, i.e., the anaerobic threshold, can be used as an index of the oxidative capacity of the muscle. Furthermore, the anaerobic threshold can be accurately determined noninvasively from specific ventilatory and gas exchange measurements, since exercise-induced elevation in blood lactate is associated with a disproportionately increasing minute ventilation in relation to $\dot{V}_{O_2}$ during incremental work. Noninvasive assessment of the anaerobic threshold before and after 10 days of bed rest in 10 subjects suggested that blood lactate accumulation occurred at a lower absolute (from 93 to 65 watts) and relative (52–42% of $\dot{V}_{O_2\ max}$) work rate following deconditioning (Convertino *et al.*, 1986b). These data are consistent with earlier observations that blood lactate levels during exercise following bed rest were greater than pre-bed-rest levels at the same work rates (Saltin *et al.*, 1968), suggesting that the rate of oxidation of pyruvate within the muscle has decreased with bed rest. The mechanism is not clear but may involve a reduction in the activities of oxidative enzymes. However, significant correlation coefficients between the change in anaerobic threshold and blood and plasma volumes (.73 and .80, respectively) following bed rest may suggest that limited oxygen transport to the muscle is an important factor.

Unfortunately, there are few available data on biochemical changes in blood or at the cellular level that would be necessary to determine alterations in the metabolic capacity of the muscle following inactivity. Muscle biopsies were obtained from three cyclists and four runners after intervals of detraining of up to 12 weeks (Chi *et al.*, 1983). During detraining, the oxidative enzymes citrate synthase, malate dehydrogenase, β-hydroxyacyl-CoA dehydrogenase, succinate dehydrogenase, and β-hydroxybutyrate dehydrogenase were decreased in both slow-oxidative (SO) and fast-oxidative-glycolytic (FOG) muscle fibers, whereas enzymes associated with anaerobic glycolysis and phosphagen utilization (phosphorylase, lactate dehydrogenase, and adenylate kinase) were increased. Rifenberick and co-workers (1973) reported that disuse atrophy induced by immobilization of the hind limb in rats reduced the yield of mitochondrial protein that was associated with decreased activities of cytochrome oxidase, monoamine oxidase, and malate dehydrogenase. Following 8 weeks of detraining in 24 men, Orlander and associates (1977) found a significant decrease in 3–hydroxy-CoA dehydrogenase, representing intact mitochrondrial enzyme capacity for fatty acid β oxidation, but they observed no changes in the activities of phosphofructokinase, lactate dehydrogenase, citrate synthase, and cytochrome oxidase. These findings suggest that prolonged muscle inactivity induces a significant reduction in aerobic metabolism at the cellular level, which in turn could contribute to the decreased $\dot{V}_{O_2 \, max}$ and physical work capacity observed after bed rest by diminishing the rate at which the muscle could utilize oxygen and thus provide the required chemical energy (ATP to ADP) to muscles during maximal work. However, further studies are needed to examine the time course of enzyme changes and to correlate the findings with $\dot{V}_{O_2 \, max}$ changes to better define the role of cellular metabolic rate as a contributing factor.

Substrate availability to the working muscle may also serve as a mechanism for limiting work output following bed rest. Peripheral glucose uptake is dramatically reduced following bed rest (Dolkas and Greenleaf, 1977; Greenleaf and Kozlowski, 1983), which may limit the availability and utilization of carbohydrates during exercise. Paradoxically, a greater proportion of the energy transfer during exercise following bed rest appears to depend on greater carbohydrate metabolism, as suggested by the observed elevations in the respiratory exchange ratio and blood lactate levels. Thus, greater utilization of muscle glycogen stores, with less utilization of free fatty acids, might be required to sustain a work task following bed rest, particularly during the performance of moderate to heavy work. Further studies designed to examine the effects of inactivity on muscle glycogen stores and fat metabolism (mobilization, uptake, utilization) are necessary to better define the limitation of substrate availability and utilization as a contributing factor to the resulting exercise intolerance.

IV. ENVIRONMENTAL FACTORS AFFECTING EXERCISE PERFORMANCE AFTER BED REST

The absence of regular daily physical activity is one explanation for the significant decrease in the $\dot{V}_{O_2 \, max}$ and physical working capacity of an individual confined to bed rest. However, an alternative explanation is that the capability of the cardiovascular system to adequately respond during stresses such as exercise is diminished in the absence of normally occurring changes in hydrostatic pressure loading and venous pooling, i.e., reduction of the regular cardiovascular orthostatic stress produced by a natural standing position.

A. Physical Activity during Bed Rest

In several studies, specific exercise regimens have been used during bed rest to provide an effective countermeasure against the deconditioning effects of immobilization. Since prolonged chair rest resulted in changes in orthostatic and exercise tolerance similar to those induced by continuous supine bed rest, it has been suggested that simple inactivity and restriction of body movement in the presence of a normal gravitational environment is sufficient to result in deconditioning (Lamb, 1964; Lamb et al., 1964a,b, 1965). Furthermore, Birkhead and associates (1963a, 1964b) reported that daily sitting exercise for 1 hr during 24 days of bed rest maintained physical working capacity but not orthostatic (tilt) tolerance, whereas quiet sitting for 8 hr daily during 30 days of bed rest maintained tilt tolerance but not physical work capacity. In another study (Hoche and Graybiel, 1974), 12 subjects performed 2-hr treadmill exercise at an intensity of 15–30% of $\dot{V}_{O_2 \, max}$ twice daily at one-half gravity to prevent loss of exercise capacity and of orthostatic tolerance following 14 days of combined water immersion and bed rest. When compared with a control group that did not undergo exercise, the decline in exercise capacity (as measured by test duration on a treadmill) and plasma volume was twofold less in the exercise group whereas the loss of orthostatic tolerance was similar in both groups. These data indicated that physiological adaptations to the stresses of orthostasis and exercise are separable and that not all are attributable to inactivity per se. However, since the exercise treatments used in these latter studies were performed with the subjects exposed to the upright position, it remains difficult to ascertain whether a reduction in work capacity following bed rest inactivity results specifically from a lack of regular muscular activity or from reduced exposure to the upright posture. If the absence of daily physical activity is the primary cause of the reduced exercise capacity that accompanies bed rest, then exercise training performed in the supine position should be a useful countermeasure against reduction in $\dot{V}_{O_2 \, max}$ and thus exercise performance. However, the effectiveness of such in-bed training probably depends on the qualities of exercise such as intensity and mode, yet data from a significant number of investigators have indicated that isotonic exercise alone has been only partially effective in restoring exercise tolerance following prolonged bed rest. Chase and co-workers (1966) reported a small increase in $\dot{V}_{O_2 \, max}$ from 3.19 to 3.42 liter/min (+7.2%) with bicycle ergometer training during 15 days of bed rest but found a significantly greater elevation of

$\dot{V}_{O_2 \text{ max}}$ from 2.96 to 3.42 liter/min (+51.5%) when the training consisted of supine trampoline work for 30 min daily. These data suggest that the sharply alternating accelerations may have induced specific responses of the cardiovascular system to provide significant remedial effects and that gravitational exposure is an important factor in the mechanism of bed-rest-induced reductions in $\dot{V}_{O_2 \text{ max}}$. However, various exercise-training regimens that have been used as remedial treatments have been only partially effective or have failed to restore $\dot{V}_{O_2 \text{ max}}$ to pre-bed-rest values. Submaximal $\dot{V}_{O_2}$, heart rate, gas exchange, and muscle ataxia during treadmill exercise following 30 days of bed rest were similar in subjects who underwent in-bed isotonic exercise training and in nonexercised controls (Stepantsov et al., 1972). However, the oxygen deficit incurred during exercise following 20 days of bed rest, which is an index of anaerobic metabolism during exercise, was less in subjects who underwent bicycle ergometer exercise during bed rest compared to nonexercising control subjects (Iseyev and Katkovskiy, 1968). Despite the use of regular isotonic supine bicycle ergometer exercise for 60 min daily, Miller and co-workers (1965) reported a decrease in $\dot{V}_{O_2 \text{ max}}$ from 2.91 to 2.28 liter/min (-21.6%) following 28 days of bed rest. They attributed the lack of exercise training effect on the minimal exercise intensity used during bed rest ($\dot{V}_{O_2}$ of 813 ml/min). Meehan and associates (1966) observed a $\dot{V}_{O_2 \text{ max}}$ reduction from 3.75 to 3.04 liters per min (-18.9%) in 14 subjects confined to 28 days of bed rest, despite combinations of isotonic exercise and pressure breathing exposures. Birkhead and associates (1963a, 1964b) reported that 1 hr of daily supine exercise during bed rest resulted in a $\dot{V}_{O_2 \text{ max}}$ decrease of 6.6% compared with a 12.6% $\dot{V}_{O_2 \text{ max}}$ reduction after 42 days of continuous bed rest without exercise.

Constant muscular resistance produced by specially designed suits has also been used to prevent the decrease in exercise tolerance induced by prolonged bed rest. These garments provided a work load on the musculoskeletal system of both the upper and lower body approximating that experienced when the subject is exposed to the gravity vector along the axis of the body. Despite the fact that the results from these studies indicated no difference in the cardiovascular response to orthostasis following bed rest (Lancaster and Triebwasser, 1971; Vogt et al., 1967), the cardiovascular and metabolic responses to exercise have not been adequately reported. During 62 days of bed rest, exercise using springs and large rubber bands, which produced a workload similar to that produced from wearing the exercise garments, increased isometric and isotonic endurance as well as muscle strength (Cherepakhin, 1968). This type of resistive exercise garment, therefore, may be partially beneficial in maintaining exercise tolerance following bed rest.

Since the reduction in exercise performance following muscular inactivity has been associated with reduced electrical (electromyograph) activity (Booth, 1982), several Soviet scientists have employed electrical stimulation to muscle groups during bed rest as a possible preventive measure. Daily electrical stimulation of muscles during 30 days of bed rest produced faster recovery of the cardiovascular responses to exercise and passive standing compared with non-stimulated control subjects (Beregovkin and Kalinichenko, 1974). Kakurin and associates (1975) reported that muscle atrophic processes, as well as reductions in muscle tone and $\dot{V}_{O_2 \ max}$ observed in control subjects, were prevented by electrical stimulation and daily exercise during bed rest. It is difficult, however, to determine how much of this effect resulted from the exercise or from the direct electrical stimulation.

Although the use of isometric exercise during bed rest has not been entirely successful in eliminating the deconditioning effects of bed confinement on exercise performance (Cardus et al., 1965; Miller et al., 1964), it appears to be more effective than isotonic training. The use of a combination of isotonic–isometric exercise during bed rest can maintain body protein if the exercise intensity is 400 kcal/day (Syzrantsev, 1967). This maintenance of muscle protein may be associated with the observed retention of muscle tone, muscular strength, and exercise tolerance (Kakurin et al., 1970; Katkovskiy et al., 1969). The data of Stremel and co-workers (1976) indicated that exercise training alone during bed rest reduces but does not eliminate the decrease in $\dot{V}_{O_2 \ max}$. The $\dot{V}_{O_2 \ max}$ was reduced in subjects who underwent daily treatments of 60-min isometric and isotonic leg exercise during 14 days of bed rest, but this change in aerobic capacity was significantly less compared with the 12.3% decrement in $\dot{V}_{O_2 \ max}$ in subjects exposed to bed rest without any remedial treatment, as shown in Table III.

Although neither isometric nor isotonic exercise regimens produced remedial effects on bed rest deconditioning as measured by centrifugation tolerance, muscle ataxia, and body balance, both forms of exercise appeared to reduce the loss of plasma volume during bed rest (Greenleaf et al., 1977; Haines, 1974). A surprising finding from these studies was that isometric leg exercise during bed rest resulted in a $\dot{V}_{O_2 \ max}$ decrease of only 4.8% compared with a 9.2% reduction after isotonic leg exercise training. These investigators suggested that an additional positive hydrostatic effect of routine exposure to gravity may be necessary to restore $\dot{V}_{O_2 \ max}$ to ambulatory control levels and that perhaps the greater protective effect of isometric-type exercise on $\dot{V}_{O_2 \ max}$ resulted from a greater hydrostatic component from the static muscular contractions, which more closely simulated venous system fluid shifts that normally occur in standing. These data are consistent with previous observations that regular physical activity during bed rest is most effective when exposure to gravity is added (Birkhead et al., Graybiel, 1974).

B. Orthostatic Factors and Exercise after Bed Rest

The importance of regular exposure to the orthostatic stress of gravity on the cardiovascular system is further supported by the data of Convertino and co-

TABLE III

Mean Changes in Maximal Oxygen Uptake ($\dot{V}_{O_2 max}$) before and after Bed Rest, with Various Remedial Treatments during the Bed Confinement Period

Study	No. of male subjects	Age (years)	Bed rest duration (days)	Remedial treatment Mode	Remedial treatment min/day	$\dot{V}_{O_2 max}$ (liter/min) Before	$\dot{V}_{O_2 max}$ (liter/min) After	$\dot{V}_{O_2 max}$ (liter/min) % Δ
Stremel et al. (1976)	7	19–22	14	None	—	3.83	3.36	−12.3
Stremel et al. (1976)	7	19–22	14	Isotonics	60	3.80	3.45	−9.2
Stremel et al. (1976)	7	19–22	14	Isometrics	60	3.77	3.59	−4.8
Convertino et al. (1982c)	4	20–26	14	None	—	3.86	3.32	−14.0
Convertino et al. (1982c)	4	19–22	14	Venous pooling	210	3.29	3.13	−4.9

workers (1982c) (Table III) in comparison with the results of Stremel and co-workers (1976). In the study by Convertino *et al.*, 4 subjects, while remaining recumbent in bed for 14 days, received daily 210-min treatments with a reverse-gradient garment especially designed to produce the same degree of venous pooling experienced by a person who sits, stands, and walks (Annis, 1974). It was concluded that the reduction in hydrostatic pressure-induced pooling (i.e., regular cardiovascular exposure to gravity) during bed rest significantly contributes to a reduction in exercise performance as measured by $\dot{V}_{O_2\ max}$, since the decrease in $\dot{V}_{O_2\ max}$ with venous pooling treatment was only 4.9% compared with a 14.0% loss in nontreated subjects. The minimal 4.9% decrease in $\dot{V}_{O_2\ max}$ with venous pooling treatment is closely comparable to the 4.8% $\dot{V}_{O_2\ max}$ reduction observed in subjects treated with isometric leg exercise in the study by Stremel and associates (1976). These beneficial effects of minimizing cardiovascular deconditioning by manipulating hydrostatic-pressure-induced blood pooling have been substantiated by other investigators (Buyanov *et al.*, 1967; Miller *et al.*, 1964).

Since the regular presence of minimal gravitational effects on the cardiovascular system apparently provides significant protection against a bed-rest-induced decrease in $\dot{V}_{O_2\ max}$, the reduction in exercise performance may result primarily from an inability of cardiovascular mechanisms to compensate adequately during the orthostatic stress of resuming the upright position following prolonged bed rest. The effects of orthostatic stress on exercise performance after bed rest can be more precisely determined by comparing cardiorespiratory responses to upright and supine exercise before and after bed rest. The effect of decreased physical activity associated with bed rest can be isolated by measuring changes in $\dot{V}_{O_2\ max}$ during supine exercise, whereas $\dot{V}_{O_2\ max}$ changes with upright exercise define the additional contribution of body position and hydrostatic gradient changes (i.e., orthostatic stress) within the vascular system to performance during exercise. If orthostatic stress is a significant factor contributing to the reduction in exercise performance, then the changes in the cardiorespiratory response to exercise performance in the upright position following bed rest should be greater than those changes during supine exercise, because of the added stress of orthostatic hypotension (Cooper and Ord, 1968).

The data in Table IV summarize the changes in $\dot{V}_{O_2\ max}$ reported in eight different studies using supine and upright exercise tests to assess the effect of bed rest deconditioning on exercise performance. The average decrease in $\dot{V}_{O_2\ max}$ reported in the studies using supine exercise was 9.1% compared with 18.4% observed from the combined studies including upright exercise testing. In a study specifically designed to determine the effect of orthostatic stress on exercise performance following bed rest, Convertino and associates (1982b) measured the changes in both supine and upright $\dot{V}_{O_2}$ following 10 days of bed rest in 12 middle-aged (45–55 years) subjects. The principal finding of this study was a

TABLE IV

Mean Changes in Supine (SUP) and Upright (UP) Maximal Oxygen Uptake ($\dot{V}_{O_2\,max}$) before and after Various Durations of Bed Rest

Study	N	Sex	Age (years)	Bed rest duration (days)	Exercise position	$V_{O_2,max}$ (liters/min) Before	After	$\% \Delta$
Stremel et al. (1976)	7	M	19–22	14	SUP	3.83	3.36	−12.3
Convertino et al. (1977)	15	M	19–23	14	SUP	3.52	3.20	−9.1
Convertino et al. (1977)	8	F	23–34	17	SUP	2.06	1.86	−9.7
Convertino et al. (1986a)	15	M	45–65	10	SUP	2.78	2.52	−9.4
Convertino et al. (1986a)	15	F	45–65	10	SUP	1.65	1.54	−6.7
Convertino et al. (1982b)	12	M	45–55	10	SUP	2.02	1.87	−7.2
Convertino et al. (1982b)	12	M	45–55	10	UP	2.14	1.78	−16.9
Convertino et al. (1982c)	4	M	20–26	15	UP	3.86	3.32	−14.0
Kakurin et al. (1966)	4	M	22–24	20	UP	3.10	2.70	−12.9
Miller et al. (1965)	6	M	18–21	28	UP	2.91	2.28	−21.6
Saltin et al. (1968)	5	M	19–21	20	UP	3.30	2.43	−26.4

significantly greater decline in $\dot{V}_{O_2\,max}$ during upright than during supine exercise after bed rest, i.e., approximately 17% versus 7% (Table IV). The larger reduction in the rate of oxygen transport and/or utilization in upright compared with supine exercise following bed rest is associated with a greater reduction in stroke volume and cardiac output (Hung et al., 1983; Saltin et al., 1968), higher heart rate and heart-rate–blood-pressure product (Convertino et al., 1981, 1982a,b; Hung et al., 1983), and larger recruitment of anaerobic energy supply, as suggested by lower submaximal and maximal oxygen uptake associated with greater submaximal oxygen deficit, oxygen debt, $\dot{V}_{O_2}$ time constant, and respiratory quotient (Convertino et al., 1981, 1982a, 1984, 1986b; Convertino and Sandler, 1982). Thus, based on the results of changes in supine and upright $\dot{V}_{O_2\,max}$ following bed rest, it appears that the absence of regular physical activity and venous system fluid shifts induced by orthostatic factors (i.e., gravity) contribute nearly equally to the bed-rest-induced reduction in $\dot{V}_{O_2\,max}$.

V. RECOVERY FROM BED REST INACTIVITY

A. Role of Exercise Training

Although immediate post-bed-rest physiological limitations to exercise are a primary concern with regard to work performance, it is also necessary to consider the effects of a long-term recovery rate as a factor limiting the resumption of high-intensity physical activity. Since cardiovascular and metabolic changes induced by bed rest may require 2–6 weeks to return to pre-bed-rest control levels (Deitrick *et al.*, 1948; Kakurin *et al.*, 1963; Lipman et al., 1969), it is essential to critically examine factors that may enhance or impede recovery from bed rest deconditioning.

The classic bed rest study of Saltin and co-workers (1968) is often cited as evidence favoring the use of exercise conditioning programs as an effective technique for enhancing the recovery from the deleterious effects of bed rest on exercise performance. In the 3 habitually sedentary subjects studied by these investigators, $\dot{V}_{O_2 \; max}$ levels (which were reduced by 31% following bed rest) were restored within 10–14 days of recovery from bed rest and continued to increase by 36% above pre-bed-rest levels at 60 days of recovery. However, the 2 habitually active subjects in this study required 30–40 days of physical activity to restore $\dot{V}_{O_2 \; max}$ values to pre-bed rest-levels. These data suggest that exercise training following bed rest deconditioning can enhance the recovery rate of $\dot{V}_{O_2 \; max}$ and physical working capacity in more sedentary individuals, but this is not the case in already trained individuals.

One potential problem in the interpretation of the data of Saltin and co-workers is that the study was not designed to distinguish between the effects of exercise conditioning and those of resuming usual activities in the normal upright position; all 5 subjects in the study underwent exercise training. However, in a recent study by DeBusk and associates (1983) of 12 healthy middle-aged men (45–55 years) who had been at bed rest for 10 days, 6 were randomly assigned to perform individually prescribed physical exercise daily for 60 days after bed rest (exercise group), and 6 simply resumed their customary activities (control group). Despite a significantly greater increase in $\dot{V}_{O_2 \; max}$ in the exercise group at 60 days compared with the control group, the $\dot{V}_{O_2 \; max}$ and physical working capacity in both groups returned to pre-bed-rest levels by 30 days after bed rest, and this was accompanied by significant and similar increases in resting left ventricular end-diastolic and stroke volumes in both groups. It was concluded that simple resumption of usual physical activities after bed rest was as effective as formal exercise conditioning in restoring the functional capacity to pre-bed-rest levels. These results are further supported by more recent data demonstrating that (a) a randomized trial of in-hospital exercise conditioning in postinfarction patients did not increase treadmill performance 10 days after their acute event (Sivarajan *et al.*, 1981) and (b) pre-bed-rest $\dot{V}_{O_2 \; max}$ values were restored by 14 days of recovery from a 10-day bed rest period in 9 healthy middle-aged men (35–50 years) who merely resumed normal daily activities with no daily exercise (Convertino *et al.*, 1985). It should be noted, however, that older, less fit individuals were examined in these latter three studies that appear to support normal resumption of daily activities in the upright posture as the primary stimulus to recovery from bed rest deconditioning, in contrast to the 5 young (19–20 years), relatively fit subjects used in the Saltin *et al.*, (1968) study. Thus, the effectiveness of resuming normal upright activities following bed rest in enhancing recovery from the deconditioning effects may depend on the age and fitness level of the population involved.

B. Effect of Repeated Bed Rest

The return of the cardiovascular state to a pre-exposed level after bed rest or spaceflight has varied and generally depends on the duration of exposure and the use of countermeasures (Gazenko *et al.*, 1981; Hyatt *et al.*, 1973; Michel *et al.*, 1979; Vetter *et al.*, 1971). It also depends on the variables measured, with plasma volume changes usually returning rapidly within 28–48 hr (Leach, 1979). Other indicators of integrative or nervous system control (heart rate, systolic time intervals) or adaptive responses to inactivity (muscle or bone loss or both) require up to a week, a month, or more to return to pre-bed-rest values (Hyatt *et al.*, 1973; Sandler, 1983). A large number of bed rest studies lasting 7–14 days (see Tables I and III) have required the same length of time spent in upright activity to restore exercise tolerance to pre-bed-rest levels. Little or no information is available on the time required to recover from longer periods of bed rest. Furthermore, data are complicated by heavy use of countermeasures.

However, some quantitative insight into the nature of the recovery process has come from studies in which repeated bed rest exposures were separated by periods of upright activity. Stremel and associates (1976) studied 7 young, athletic males before and after three consecutive 2-week bed rest exposures separated by 3-week recovery periods in which the subjects exercised for 60 min per day on bicycle ergometers at 50% of their $\dot{V}_{O_2 \text{ max}}$. They found that $\dot{V}_{O_2 \text{ max}}$, maximal minute ventilation, maximal heart rate, and exercise tolerance time were the same at the end of recovery from the second bed rest period compared with that of the first.

Earlier, Cardus (1966) had reported that heart rate during 40, 80, and 120 watts of bicycle ergometer exercise was increased and that $\dot{V}_{O_2}$ at a heart rate of 160 beats per minute was decreased following each of three 10-day bed rest periods separated by 3-week periods of normal daily activities. The magnitude of the bed-rest-induced changes in exercise heart rate and $\dot{V}_{O_2}$ were similar following each deconditioning period. These changes in cardiovascular responses to exercise persisted after 1 week of recovery but returned to pre-bed-rest control values by the second or third week of recovery.

In a more recent study (Convertino *et al.*, 1985), 7 middle-aged men (35–50 years) underwent two 10-day bed rest periods separated by 14 days of recovery in which they followed regular daily activities without formal exercise. The $\dot{V}_{O_2 \text{ max}}$, anaerobic threshold, and oxygen pulse decreased, whereas the submaximal and maximal heart rates, diastolic pressures, and rate–pressure products increased following bed rest. Although the submaximal and maximal $\dot{V}_{O_2}$ values returned to pre-bed-rest control levels following 14 days of recovery, submaximal and maximal heart rates, diastolic pressures, and rate–pressure products remained significantly elevated following recovery from the second bed rest exposure. With regard to metabolism and muscular work performance, two weeks of minimal activity are adequate for complete recovery from bed rest deconditioning, and repeated exposure can be safely tolerated. However, available data suggest that altered physiological effects may persist much longer than previously realized. Furthermore, there appears to be a cumulative effect following a second bed rest exposure, particularly in middle-aged subjects, which is best manifested by indices of increased myocardial work during exercise.

VI. PRACTICAL APPLICATIONS

A. Clinical Implications

The available exercise data following prolonged bed rest have important implications for the in-hospital rehabilitation of patients confined to bed, particularly myocardial infarction (MI) patients. In evaluating the reduced exercise tolerance of patients recovering from MI and other diseases, it is difficult to distinguish the role of cardiac damage and other contributing diseases from that of bed rest deconditioning. This distinction has now become possible to assess, due to the large accumulated data base on the effects of prolonged bed rest on healthy subjects (Sandler, 1980). Although most of the early investigations using exercise involved young, healthy subjects (18–25 years), the data from more recent studies have been collected from subjects ages 45–65 years (Convertino *et al.*, 1982a,b, 1984, 1985, 1986a,b; DeBusk *et al.*, 1983; Hung *et al.*, 1983) and are of particular interest, since cardiovascular and metabolic changes following bed rest were examined in otherwise healthy subjects but of an age range when MI and other debilitating diseases are common. In addition, the duration of bed rest was similar to that usually prescribed for patients recovering from acute MI. It is apparent from the results of these studies that orthostatic intolerance induced by bed rest deconditioning can contribute to the reduced cardiorespiratory capacity observed after MI, independent of cardiac damage. Additionally, an augmented heart rate and heart-rate–blood-pressure product during submaximal work was observed during testing of these normal subjects (Hung *et al.*, 1983). Since heart rate and heart-rate–blood-pressure product are important determinants of myocardial oxygen demands, the presence of a deconditioned state would result in a need for greater cardiac work (less efficient), the increased myocardial oxygen demand being deleterious, or even potentially hazardous, to patients with chronic ischemic heart disease, particularly those with limited reserve (Gobel *et al.*, 1978). The use of appropriate countermeasures could minimize or eliminate these problems by returning the upright post-bed-rest heart-rate–blood-pressure product to or towards pre-bed-rest levels. In the short term, this approach could decrease the risk of angina pectoris and clinical coronary events during orthostatic stress in patients with significantly restricted coronary blood flow. Although strategies to limit the decrease in oxygen transport capacity after MI have emphasized low-level supine or upright exercise training, currently available data from various bed rest and clinical studies (Convertino, 1983a; DeBusk *et al.*, 1983; Hung *et al.*, 1983; Sivarajan *et al.*, 1981) suggest that simple exposure to gravitational stress (sitting upright, dangling of legs at bedside) helps accomplish this purpose as well. Such benefits have been already discussed in Chapter 2 and are due to a decrease in heart size itself when the subject is upright, leading to a decrease in myocardial oxygen demand per beat on this basis alone (Coe, 1954). This underscores the importance of intermittent sitting or early ambulation during the hospital confinement period in preventing or limiting unnecessary physical debility after MI. Chair rest to treat MI has been used empirically for almost five decades. Thus, it appears that deterioration in exercise performance resulting from bed rest may be largely obviated by regular exposure to gravitational stress, and formal programs of in-hospital rehabilitation emphasizing exercise training, therefore, may not be necessary for this purpose (DeBusk *et al.*, 1983).

B. Implications for Sedentary Life-Styles

The study of biological adaptation to acute and chronic exercise has emphasized the role of physical training that allows individuals to increase their work capacity. It is equally important, however, that we understand the consequences of the detraining, or reduced physical activity, associated with more sedentary life-styles that are adopted because of working and social habits, as well as the aging process. Although the bed rest model is not completely analogous to sedentary living habits, many of the physioloigical changes that occur during bed rest exposure are qualitatively similar to those observed during detraining or aging. Thus, data obtained from bed rest studies may be helpful in providing a relatively short-term research tool to examine the deleterious effects of sedentary living on exercise performance.

The human body adapts itself to sedentary living habits. If the body is regularly exposed to little or no exercise stress, there is a dramatic reduction in the physical work capacity, which is associated with cardiovascular and metabolic changes similar to those observed during bed rest. These physiological and physical changes depend on the length of the deconditioning period as well as on the initial fitness level of the individual. Thus, sedentary living has more profound and measureable effects on a former athlete than on an individual who has always been sedentary, regardless of age.

Data from bed rest studies strongly suggest that a primary mechanism for maintaining functional work capacity is the regular exposure to the upright posture during daily activities. Thus, even if an individual is limited by the time he or she can spend in high-intensity physical activities, providing regular upright activity such as standing or walking, or both, is essential for maintaining an adequate and healthy physiological tolerance to exercise. This has been demonstrated particularly in middle-aged or older populations (DeBusk et al., 1983). As demonstrated by various bed rest studies, regular minimal exposure to upright physical activity may not stimulate a high capacity for work output, but it may deter the onset of deconditioning resulting from sedentary life-styles as well as from the aging process.

C. Implications for Aerospace Medicine

Human physiology is extremely adaptive to its external environment. Indeed, upon exposure to the absence of gravity during spaceflight, the organ systems of our bodies adapt at a surprisingly rapid rate to meet the demands of the new environment. Thornton (1981) pointed out that with respect to spaceflight, "adaptation should be distinguished from deconditioning, which carries the connotation of an undesirable or unhealthy state. In contrast, the adaptation to weightlessness appears to be a stable, healthy state." No in-flight, adverse physiological effects have occurred at rest to date, yet increased arrhythmias during exercise and excessive heart rates during extravehicular activity have been reported (Smith et al., 1977). Significant clinical cardiovascular problems occur only when astronauts or cosmonauts return to earth. Since plans for a space station remain many years in the future (at least beyond the year 2000), each space traveler must eventually return once again to earth. Similarly, normal social function on earth cannot be accomplished while a person is bedridden or immobilized, since it requires upright body posture and the ability to locomote. Thus, even though all of the physiological changes that occur during simulated or actual spaceflight may be considered appropriate body adaptations to a new environment, the desirability of prohibiting these natural adaptations must be seriously addressed in relation to an individual's future health status and the ability to perform or to survive when the person returns to normal activity on earth.

Bed rest has been used as an effective model for simulating many of the physiological changes induced by spaceflight, particularly with regard to cardiovascular and metabolic responses to orthostasis and exercise, as well as bone and muscle changes (Convertino et al., 1981; Katkovskiy and Pometov, 1976; Rummel et al., 1973; Sandler, 1976, 1980). Therefore, with the emergence of the space shuttle program and plans for a future space station with subsequently more frequent or longer-duration exposures to weightlessness, the effect of prolonged bed rest confinement on exercise performance and the use of exercise programs during bed rest as a countermeasure to the deconditioning response is relevant to the possible solutions of physiological problems incurred by astronauts during spaceflight.

Based on the data obtained from various bed rest studies, it is evident that the reduction in physical work capacity during and following prolonged spaceflight is a net result of the absence of gravity combined with reduced physical activity. Exercise training regimens during simulated weightlessness have been only partially effective in restoring the physical work capacity of individuals. The most effective or potentially effective programs have employed exercise that had qualities of gravitational stress on muscle or the cardiovascular system or both, such as isometrics or the use of a trampoline (Chase *et al.*, 1966). However, since the biological adaptations to the nongravity environment of space appear to be most natural, it may not be necessary to maintain the same level of physical fitness during flight as that required on earth. In fact, except for extravehicular activities, which have some specific strength and endurance requirements, spaceflight presents no extraordinary physical demands—actually, it is a very sedentary environment. Therefore, the data from bed rest models indicate that appropriate exercise programs can be implemented during spaceflight to provide the astronaut with fitness to perform adequately the physical work tasks required during spaceflight. These may be composed of a combination of arm and/or leg static and dynamic exercises, as well as exposure to constant muscle resistance through the use of special spring-loaded suits. These countermeasures are currently being used during spaceflight with varying degrees of success (Sandler, 1980).

A major problem with regard to exercise tolerance post-flight is the inability of the astronaut to maintain "normal" cardiovascular and metabolic function during exercise on return to earth. The subsequent early limitation in exercise performance appears to be largely an orthostasis problem associated with loss of intravascular volume and later a loss in muscle mass. Although reduction in body fluids and blood volume may be advantageous for effective adaptation during weightlessness, it is beneficial for these volumes, as well as other physiological functions, to be restored just prior to returning to the 1-g environment. Simple fluid loading may not be completely effective, as shown by echocardiographic measurements of end-diastolic left ventricular volume in astronauts or cosmonauts postflight (Sandler *et al.*, 1985). Oral or intravenous isotonic saline taken several hours before reentry may not remain intravascularly since it can move rapidly (within 10–20 min) to the extravascular spaces. Data from past spaceflights and bed rest studies do not strongly support the long-term use of exercise alone for achieving the restoration of 1-g functions during weightlessness. The use of special suits designed to induce LBNP during exercise, however, may be effective when used 24–48 hours prior to reentry. Furthermore, short-term, high-intensity exercise can induce large increases in plasma and blood volume, with associated improvement in the cardiovascular reflex adjustments to orthostasis (Convertino *et al.*, 1984), suggesting that exercise alone may be effective in restoring preflight cardiovascular functioning during muscular work if the intensity and duration of the in-flight exercise regimens are high enough. Further research is needed to identify the optimum exercise stimulus for the preentry recovery from spaceflight.

VII. SUMMARY

The exercise response after bed rest inactivity is a reduction in the physical work capacity and is manifested by significant decreases in oxygen uptake. The magnitude of decrease in maximal oxygen uptake ($\dot{V}_{O_2 \ max}$) is related to the duration of confinement and the pre-bed-rest level of aerobic fitness; these relationships are relatively independent of age or gender. The reduced exercise performance and $\dot{V}_{O_2 \ max}$ following bed rest are associated with various physiological adaptations including reductions in blood volume, submaximal and maximal stroke volume, maximal cardiac output, skeletal muscle tone and strength, and aerobic enzyme capacities, as well as increases in venous compliance and submaximal and maximal heart rate. This reduction in physiological capacity can be partially restored by specific countermeasures that provide regular muscular activity or orthostatic stress or both during the bed rest exposure. The understanding of these physiological and physical responses to exercise following bed rest inactivity has important implications for the solution to safety and health problems that arise in clinical medicine, aerospace medicine, sedentary living, and aging.

REFERENCES

Annis, J.F. (1974). *Aerosp. Med. Assoc. Preprints*, 96-97.

Astrand, I. (1960). *Acta Physiol. Scand. Suppl.* **169**, 1-48.

Barcroft, J. (1934). *In*: "Features in the Architecture of Physiological Function," p. 286, Cambridge University Press, New York.

Bassey, E.J., Bennett, T., Birmingham, A.T., Fentem, P.D., Fitton, D., and Goldsmith R. (1972). *J. Physiol.* **222(1)**, 79P.

Bassey, E.J., Bennett, T., Birmingham, A.T., Fentem, P.D., Fitton, D., and Goldsmith, R. (1973). *Cardiovascular Res.* **7**, 588-592.

Beregovkin, A.V., and Kalinichenko, V.V. (1974). *Space Biol. and Med.* **8(1)**, 106-112.

Bevegard, S., Holmgren, A., and Jonsson, B. (1960). *Acta Physiol. Scand.* **49**, 279-298.

Birkhead, N.C., Blizzard, J.J., Daly, J.W., Haupt, G.J., Issekutz, B., Myers, R.N., and Rodahl, K. (1963a). *In*: "Technical Report No. AMRL-TDR-63-37, pp. 1-40, Wright-Patterson Air Force Base, Ohio.

Birkhead, N.C., Haupt, G.J., Blizzard, J.J., Lachance, P.A., and Rodahl, K. (1963b). *Physiologist* **6**, 140.

Birkhead, N.C., Blizzard, J.J., Daly, J.W., Haupt, G.J., Issekutz, B., Myers, R.N., and Rodahl, K. (1964a). *In*: "Technical Report No. AMRL-TDR-64-61", pp. 1-28, Wright-Patterson Air Force Base, Dayton, Ohio.

Birkhead, N.C., Haupt, G.J., Issekutz, B., Rodahl, K. (1964b). *Am. J. Med. Sci.* **247**, 243.

Blomqvist, G., Mitchell, H.J., and Saltin, B. (1971). *In*: "NASA SP-269, Hypogravic and Hypodynamic Environments" (R.H. Murray and M. McCally eds.), pp. 171-176. Washington, DC.

Booth, F.W. (1977). *J. Appl. Physiol.: Resp. Environ.* **43**, 656-661.

Booth, F.W. (1982). *J. Appl. Physiol.: Resp. Environ. Exercise Physiol.* **52**, 1113-1118.

Booth, F.W., and Seider, M.J. (1979). *J. Appl. Physiol.: Resp. Environ. Exercise Physiol.* **47**, 974-977.

Buderer, M.C., Owen, C.A., Rummel, J.A., Sawin, C.F., Schachter, P. (1976a). *In*: "Report of 14-Day Bedrest Simulation of Skylab (NASA-CR-147758)", (P.C. Johnson and C. Mitchell, eds.), Methodist Hospital, Houston, Texas.

Buderer, M.C., Rummel, J.A., Michael, E.L., Maulden, D.C., and Sawin, C.F. (1976b). *Aviat. Space Environ. Med.* **47**, 365-372.

Buyanov, P.V., Beregovkin, A.V., and Pisarenko, N.V. (1967). *Space Biol. Med.* **1(1)**, 95-99.

Cardus, D. (1966). *Aerosp. Med.* **37**, 993-999.

Cardus, D., Spencer, W.A., Vallbona, C., and Vogt, F.B. (1965). *In*: "Cardiac and Ventilatory Response to the Bicycle Ergometer Test (NASA-CR-177)", Texas Institute for Rehabilitation and Research, Houston, Texas.

Chase, G.A., Grave, C., and Rowell, L.B. (1966). *Aerosp. Med.* **37**, 1232-1238.

Cherepakhin, M.A. (1968). *Space Biol. Med.* **2(1)**, 52-59.

Chi, M.M.-Y., Hintz, C.S., Coyle, E.F., Martin, W.H., Ivy, J.L., Nemeth, P.M., Holloszy, J.O., and Lowry, O.H. (1983). *Am. J. Physiol.* **244**, C276-C287.

Coe, W.S. (1954). *Ann. Int. Med.* **40**, 42-48.

Convertino, V.A. (1983a). *J. Cardiac Rehabil.* **3**, 660-663.

Convertino, V.A. (1983b). *Med. Sci. Sports Exercise* **15**, 77-82.

Convertino, V.A., and Sandler, H. (1982). *Physiologist* **25**, S159-S160.

Convertino, V.A., Stremel, R.W., Bernauer, E.M., and Greenleaf, J.E. (1977). *Acta Astronautica* **4**, 895-905.

Convertino, V.A., Bisson, R., Bates, R., Goldwater, D., and Sandler, H. (1981). *Aviat. Space Environ. Med.* **52**, 251-255.

Convertino, V.A., Goldwater, D.J., and Sandler, H. (1982a). *Aviat. Space Environ. Med.* **53**, 652-657.

Convertino, V.A., Hung, J., Goldwater, D., and DeBusk, R.F. (1982b). *Circulation* **65**, 134-140.

Convertino, V.A., Sandler, H., Webb, P., and Annis, J.F. (1982c). *J. Appl. Physiol.* **52**, 1343-1348.

Convertino, V.A., Goldwater, D.J., and Sandler, H. (1984). *J. Appl. Physiol.: Resp. Environ. Exercise Physiol.* **57**, 1545-1550.

Convertino, V.A., Kirby, C.R., Karst, G.M., and Goldwater, D.J. (1985). *Aviat. Space Environ. Med.* **56**, 540-546.

Convertino, V.A., Goldwater, D.J., and Sandler, H. (1986a). *Aviat. Space Environ. Med.* **57**, 17-22.

Convertino, V.A., Karst, G.M., Kirby, C.R., and Goldwater, D.J. (1986b). *Aviat. Space Environ. Med.* **57**, 325-331.

Cooper, K.H., and Ord J.W. (1968). *Aerosp. Med.* **39**, 481-484.

DeBusk, R.F., Convertino, V.A., Hung, J., and Goldwater, D. (1983). *Circulation* **68**, 245-250.

Deitrick, J.E., Whedon, G.D., Shorr, E., Toscani, V., and Davis, V.B. (1948). *Am. J. Med.* **4**, 3-35.

Dolkas, C., and Greenleaf, J.E. (1977). *J. Appl. Physiol.* **43**, 1033-1038.

Friman, G., and Hamrin, E. (1976). *Upsala J. Med. Sci.* **81**, 79-83.

Gazenko, O.G., Genin, A.M., and Yegorov, A.D. (1981). *Acta Astronautica* **8**, 907-917.

Georgiyevskiy, V.S., Kakurin, L.I., Katkovskii, B.S., and Senkevich, Y.A. (1966). *In*: "The

Oxygen Regime of the Organism and its Regulation" (N.V. Lauer and A.Z. Kilchinskaya, eds.), pp. 181-184, Naukova Dumka, Kiev.

Gobel, F.L., Nordstrom, L.A., Nelson, R.R., Jorgensen, C.R., and Wang, Y. (1978). *Circulation* **57**, 549-556.

Greenleaf, J.E., and Kozlowski, S. (1982). *In*: Exercise and Sport Sciences Reviews, Vol. 10 (R.L. Terjung, ed.), pp. 83-119. Franklin Institute Press, Philadelphia, Pennsylvania.

Greenleaf, J.E., and Kozlowski, S. (1983). *Med. Sci. Sports Exercise* **14**, 477-480.

Greenleaf, J.E., Bernauer, E.M., Juhos, L.T., Young, H.L., Staley, R.W., and Van Beaumont (1977). *J. Appl. Physiol.:Resp. Environ. Exercise Physiol.* **42**, 59-66.

Greenleaf, J.E., Van Beaumont, W., Convertino, V.A., and Starr, J.E. (1983). *Aviat. Space Environ. Med.* **54**, 696-700.

Haines, R.F. (1974). *J. Appl. Physiol.* **36**, 323.

Hargens, A.R., Tipton, C.M., Gollnick, P.D., Mubarak, S.J., Tucker, B.J., and Akeson, W.H. (1983). *J. Appl. Physiol.: Resp. Environ. Exercise Physiol.* **54**, 1003-1009.

Hoche, J., and Graybiel, A. (1974). *Aerosp. Med.* **45**, 386-392.

Holloszy, J.O., and Booth, F.W. (1976). *Ann. Rev. Physiol.* **38**, 273-291.

Holmgren, A., and Ovenfors, C.O. (1960). *Acta Med. Scand.* **167**, 267-276.

Hung, J., Goldwater, D., Convertino, V.A., McKillip, J.H., Goris, M.L., and DeBusk, R.F. (1983). *Am. J. Cardiol.* **51**, 344-348.

Hyatt, K.H., Sullivan, R.W., Spears, W.R., and Vetter, W.R. (1973). "A Study of Ventricular Contractility and Other Parameters Possibly Related to Vasodepressor Syncope." U.S. Public Health Service Hospital, San Francisco. NASA Contract Report T-81035, pp. 1-75.

Iseyev, L.R., and Katkovskiy, B.S. (1968). *Space Biol. Med.* **2(4)**, 117-124.

Iseyev, L.R., and Nefedov, Y.G. (1968). *Space Biol. Med.* **2(1)**, 60-65.

Kakurin, L.I., Katkovskiy, B.S., Kozlov, A.N., and Mukharlyamov, N.M. (1963). *In*: "Aviation and Space Medicine" (V.V. Parin, ed.), pp. 192-194, Akademiya Meditsinskikh Nauk, Moscow.

Kakurin, L.I., Akhrem-Adhremovich, R.M., Vanyushina, Y.V., Varbaronov, R.A., Georgiyevskii, V.S., Kotkovskiy, B.S., Kotovskaya, A.R., Mukharlyamov, N.M., Panferova, N.Y., Pushkar, Y.T., Senkevich, Y.A., Simpura, S.F., Cherepakhin, M.A., and Shamrov, P.G. (1966). *In*: "Soviet Conference on Space Biology and Medicine", pp. 110-117, Moscow.

Kakurin, L.I., Kamkovskiy, B.S., Giorgiyevskiy, V.S., Purakhan, Yu.N., Cherenikhin, M.A., Mikhalylov, B.M., Pimukhov, B.N., and Buryikov, Ye.N. (1970). *Vopr. Kurotol. Fizioter. Lech. Fizich. Kult.* **35**, 19-24.

Kakurin, L.I., Yegorov, B.B., Il'lina, Y.I., and Cherepakhin, M.A. (1975). *Acta Astronautica* **2**, 241-246.

Katkovskiy, B.S., and Pometov, Y.D. (1971). *Space Biol. Med.* **5(3)**, 105-113.

Katkovskiy, B.S., and Pometov, Y.D. (1976). *In*: "Life Sciences and Space Research XIV", pp. 301-305, Akademie-Verlag GmbH, Berlin.

Katkovskiy, B.S., Pilysvskiy, O.A., and Smirnova, G.I. (1969). *Space Biol. Med.* **3(2)**, 77-85.

Krupina, T.N., and Tizul, A.Ya. (1971). *Zh. Nevropatol. Psikhiatr. im. S.S. Korsakova* **71**, 1611-1617.

LaFevers, E.V., Booher, C.R., Crozie, W.N., and Donaldson, J. (1977). *In*: "JSC/Methodist Hospital 28-Day Bedrest Study, Vol. II (NASA 9-14578)", (P.C. Johnson and C. Mitchell, eds.), Lyndon B. Johnson Space Center, Houston, Texas.

Lamb, L.E. (1964). *J. Am. Med. Assoc.* **188**, 27-33.

Lamb, L.E., Johnson, R.L., and Stevens, P.M. (1964a). *Aerosp. Med.* **35**, 646-649.

Lamb, L.E., Johnson, R.L., Stevens, P.M., and Welch, B.E. (1964b). *Aerosp. Med.* **35**, 420-428.

Lamb, L.E., Stevens, P.M., and Johnson, R.L. (1965). *Aerosp. Med.* **36**, 755-763.

Lancaster, M.C., and Tribwasser, J.H. (1971). *In*: "NASA SP-269, Hypogravic and Hypodynamic Environments" (R.H. Murray and M. McCally, eds.), pp. 225-248. Washington, DC.

Leach, C.S. (1979). *Acta Astronautica* **6**, 1123-1135.

Lipman, R., Ulvedal, F., Bradley, E., and Lecocq, F.R. (1969). *Physiologist* **12**, 285.

Meehan, J.P., Henry, J.P., Brunjes, S., and DeVries, J. (1966). *In*: "NASA-CR-62073", pp. 1-54. University of Southern California, Department of Physiology, Los Angeles, California.

Michel, E.L., Rummel, J.A., Sawin, C.F., Buderer, M.C., and Iem, J.D. (1979). *In*: "The Proceedings of the Skylab Life Sciences Symposium, Vol. II, (NASA-TM X-58154)" (R.S. Johnson and L.F. Dietlein, eds.), pp. 723-762. Johnson Space Center, Houston, Texas.

Miller, P.B., Hartman, B.O., Johnson, R.L., and Lamb, L.E. (1964). *Aerosp. Med.* **35**, 931-939.

Miller, P.B., Johnson, R.L, and Lamb, L.E. (1965). *Aerosp. Med.* **36**, 1077-1082.

Olree, H.D., Corbin, B., Dugger, C., and Smith, C. (1973). *In*: "NASA Technical Report No. CR-134033 (NTIS No. N73-32008/7)." Harding College, Search, Arkansas.

Orlander, J., Kiessling, K.H., Karlsson, J., and Ekblom, B. (1977). *Acta Physiol. Scand.* **101**, 351-362.

Panov, A.G., and Lobzin, V.S. (1968). *Kosmich. Biol. Med.* **2**, 59-67.

Poliner, L.R., Dehmer, G.J., Lewis, S.E., Parkey, R.W., Blomqvist, C.G., and Willerson, J.T. (1980). *Circulation* **62**, 528-534.

Rifenberick, D.H., Gamble, J.G., and Max, S.R. (1973). *Am. J. Physiol.* **225**, 1295-1299.

Robinson, B.F., Epstein, S.E., Kahler, R.L., and Braunwald, E. (1966). *Circ. Res.* **19**, 26-32.

Rowell, L.B. (1974). *Physiol. Rev.* **54**, 75-159.

Rummel, J.A., Michel, E.L., and Berry, C.A. (1973). *Aerosp. Med.* **44**, 235-238.

Saltin, B., and Rowell, L.B. (1980). *Fed. Proc.* **39**, 1506-1513.

Saltin, B., Blomqvist, G., Mitchell, J.H., Johnson, R.L., Wildenthal, K., and Chapman, C.B. (1968). *Circulation Suppl.* **7**, 1-78.

Sandler, H. (1976). *In*: "Progress in Cardiology" (P.N. Yu and J.F. Goodwin, eds.), Chapter 9, pp. 227-270. Lea and Febiger, Philadelphia, Pennsylvania.

Sandler, H. (1980). *In*: "Hearts and Heart-Like Organs, Vol. 2" (G. Bourne, ed.), Chapter 10, pp. 435-524. Academic Press, New York.

Sandler, H. (1983). *In*: "Space Physiology, Cepadues Edition", pp. 317-333, European Space Agency (ESA), Toulouse, France.

Sandler, H., Goldwater, D.J., Popp, R.L., Spaccavento, L., and Harrison, D.C. (1985). *Am. J. Cardiol.* **55(10)**, 114D.

Sivarajan, E.S., Bruce, R.A., Almes, M.J., Green, B., Belanger, L., Lindskog, B.D., Newton, K.M., and Mansfield, L.W. (1981). *N. Eng. J. Med.* **305**, 357-362.

Smith, R.F., Stanton, K., Stoop, D., Brown, D., Janusz, W., and King, P. (1977). *In* "Biomedical Results From Skylab." NASA SP-377, pp. 339-350.

Stegemann, J., Von Framing, H.-D., and Schiefeling, M. (1969). *Pflugers Archiv.* **312**, 129-138.

Stepantsov, V.I., Tikhonov, M.A., and Yermin, A.V. (1972). *Space Biol. Med.* **6(4)**, 103-109.

Stevens, P.M., Miller, P.B., Gilbert, C.A., Lynch, T.N., Johnson, R.L., and Lamb, L.E. (1966a). *Aerosp. Med.* **37**, 357-367.

Stevens, P.M., Miller, P.B., and Lynch, T.N. (1966b). *Aerosp. Med.* **37**, 466-474.

Stremel, R.W., Convertino, V.A., Bernauer, E.M., and Greenleaf, J.E. (1976). *J. Appl. Physiol.* **41**, 905-909.

Syzrantsev, Y.K. (1967). *In*: "Problems of Space Biology, Vol. 7" (V.N. Chernigovskiy, ed.), pp. 317-322. Nauka Press, Moscow.

Taylor, H.L. (1968). *Circulation* **38**, 1016-1017.

Taylor, H.L., Henschel, A., Brozek, J., and Keys, A. (1949). *J. Appl. Physiol.* **2**, 223-239.

Thornton, W. (1981). *In*: "Conference Proceedings of Spaceflight Deconditioning and Physical Fitness, NASA Publication NASW-3469" (J.F. Parker, C.S. Lewis, and D.G. Christensen, eds.), p. 13.

Trimble, R.W., and Lessard, C.S. (1970). *In*: "Contractor Report No. SAM-TR-70-56". USAF School of Aerospace Medicine, Brooks Air Force Base, Texas.

Vetter, W.R., Sullivan, R.W., and Hyatt, K.H. (1971). *Aerosp. Med. Assoc. Preprints*, pp. 56-57.

Vogt, F.B., Mack, P.B., Johnson, P.C. (1967). *Aerosp. Med.* **38**, 1134-1137.

White, P.D., Nyberg, J.W., and White, W.J. (1966). *In*: "Proceedings of the 2nd Annual Biomedical Research Conference", pp. 117-166.

Yeremin, A.V., Bazhanaov, V.V., Marishchuk, V.L., Stepantsov, V.I., and Dzhamgarov, T.T. (1969). *In*: "Problems of Space Biology, Vol 13" (A.M. Genin and P.A. Sorokin, eds.), pp. 192-199. Nauka Press, Moscow.

8

Conclusions

**HAROLD SANDLER AND
JOAN VERNIKOS**
Cardiovascular Research Office
Biomedical Research Division
National Aeronautics and Space Administration
Ames Research Center
Moffett Field, California 94035

The past several decades have seen substantial changes in the treatment of most medical problems and postsurgical procedures. Patients are no longer put to bed for prolonged periods for many diseases and following surgery. These changes have resulted from a growing awareness that inactivity and immobilization may lead to undesirable morbidity, as well as psychosocial deterioration. The greater understanding of the resulting physiological problems has come about, in large part, through findings of the many bed rest studies conducted for the U.S. and Soviet space programs and through a lesser number of studies of immobilized and paralyzed patients conducted in the clinical setting.

All physiological systems are affected by inactivity and immobilization, some more than others. The most seriously affected are the cardiovascular system, bone, and muscle. Other systems are affected but rebound fairly rapidly with return to mobility. The primary question at present is whether the decrements seen in some of these systems may cause permanent and irreversible damage. Will heart function be permanently impaired? Will bone loss eventually be replaced? Will atrophied muscle return to a healthy state? Furthermore, how can these conditions be avoided in patients who must be inactive for prolonged periods, paralyzed stroke and accident victims, paraplegics and quadriplegics, the sedentary, the aging, and space travelers exposed to inactivity and weightlessness? Medical investigators are devoting thousands of hours of research to answer these questions.

INACTIVITY: PHYSIOLOGICAL EFFECTS

I. CARDIOVASCULAR EFFECTS

Physiological findings from ground-based studies and spaceflight have demonstrated that cardiovascular responses to immobilization and inactivity on earth are similar to those seen with exposure to weightlessness. In both cases, the most striking and well documented findings are a loss of orthostatic tolerance and changes in fluid and electrolyte balance. Initial changes are due to fluid shifts to the chest like those occurring during postural change to the supine and during spaceflight. Gauer and Henry in 1976 first demonstrated that these shifts trigger intrathoracic, low-pressure mechanoreceptors to alter renal handling of water and electrolytes by inhibiting antidiuretic hormone (ADH) secretion and by decreasing the production of aldosterone, yet more recent findings cast doubt on such mechanisms as the sole explanation and may rely more on hormonal factors released from the atrial wall or on high pressure baroreceptors. With continued inactivity or weightlessness, changes continue to occur as adaptive response to the associated hypometabolic state. As study durations lengthen from weeks to months, two areas require additional, intensive study: (a) the effects of changes in central nervous system activity resulting from change in the metabolism of neurotransmitter substances and (b) the loss of cardiac muscle mass itself, associated with the general decrease in body activity and with the loss of skeletal muscle mass.

Human access to space has carried high risk and cost, and only a few long-term missions have been possible. Concerns for safety have limited both the nature and the frequency of physiological measurements, usually focusing on noninvasive techniques. Bed rest has been the next most effective alternative, which explains why so many bed rest studies have been conducted. In recent years, U.S. and Soviet investigators have been placing bed rest subjects in the head-down position. This position appears to produce cardiovascular responses more rapidly than horizontal bed rest, as well as producing the feelings of head fullness and awareness experienced by space crews. However, since far fewer studies have been conducted using the head-down method, the effectiveness, accuracy, and reproducibility of findings vis-à-vis those of horizontal bed rest have not yet been confirmed. Consequently, many studies will be needed to compare the two and to verify the findings through monitoring physiological responses during flights.

Present plans call for older and less physically fit individuals to participate in spaceflights in the future. Additional studies will be needed to determine whether weightlessness exposure will accelerate or ameliorate cardiovascular disease in such individuals. So far, male and females up to the age of 65 years have not shown diminished capability following bed rest. Future studies must concentrate on the body's adaptive mechanisms and any deleterious effects that develop over the very long term in either younger or older individuals. Findings from such studies will be needed by the space program to determine the physiological safety of humans who may spend a year or longer in space. But like many findings of space-related research, the information gained will also be applicable to health conditions of patients consigned to long-term or lifetime immobilization or inactivity, as well as to the physiological problems of aging.

Another area in which research is critically needed is in the development of methods for counteracting the physiological deterioration that occurs with inactivity, immobilization, and weightlessness. To date, none of the countermeasures considered have been either entirely effective or reliable. There is promise that pharmacologic agents may prove to be of benefit, although none as yet have done so. In the absence of such proof, research has continued on improving other techniques, such as exercise, electrostimulation, restrictive garments, and the ingestion of various fluids.

The information generated by bed rest and immersion studies, as well as animal and human spaceflights, should provide a broader understanding of the mechanisms underlying cardiovascular deconditioning and perhaps lead to a solution to these problems for benefits on earth. Some of this has already occurred, since clinical practices of early ambulation after surgery or myocardial infarction are now common practices and are having ever-wider applications in health care practices. It is also possible that the data obtained will provide useful insights into the treatment of hypertension and atherosclerotic heart disease in the clinical environment.

II. EFFECTS ON BONE

During prolonged immobilization (and weightlessness), skeletal bone structure is demineralized through the accelerated urinary excretion of calcium leading eventually to a loss of bone strength. This phenomenon is seen not only during exposure to weightlessness, but also with inactivity, bed rest studies, and prolonged clinical immobilization. Although clinical immobilization removes the weight-bearing strains from bones and joints and therefore is useful for orthopedic problems, it also results in bone demineralization over the long term. The major problem with bone demineralization from inactivity or immobilization (or spaceflight) is that we do not know whether the condition is reversible. The required carefully planned and conducted studies have not been accomplished to date. If further research proves that decrements in bone are irreversible, we may expect further changes in treatment procedures.

In general, control of bone calcium is a highly complex process dependent on the maintenance of blood calcium and phosphorus levels which are under hormonal control (mainly parathyroid hormone and vitamin D) acting through the regulation of skeletal, renal, and gastrointestinal interfaces. Excessive blood calcium leads not only to urinary losses but to related problems such as a tendency to nephrocalcinosis and nephrolithiasis, the latter of which is common in patients with spinal cord lesions, and the deposit of calcium salts in soft tissues. With continued loss of body calcium salts, a true state of osteoporosis occurs. Bone mineral loss can be evaluated by (a) measuring the urinary excretion of calcium, (b) photon beam densitometry which identifies changes as low as 2 %, (c) histological examination of bone biopsies, and (d) isotope calcium turnover studies. More traditional radiograms are not as satisfactory as the preceding techniques, because the bone mineral loss must be in the area of 40–50% before it can be detected by this means.

Numerous countermeasures have been tested in an effort to inhibit bone resorption with immobilization and spaceflight, but most have been disappointing. Remobilization of the individual, including weight-bearing and muscular activity, appears to offer the most effective means, whereas administration of dietary calcium, phosphorus, calcitonin, and diphosphanates has not proved entirely effective in counteracting hypercalciuria and osteoporosis in immobilized individuals.

The lack of knowledge concerning the reversibility of bone mineral loss and the inability thus far to determine a truly effective means of offsetting this problem indicate that much comprehensive research is still required in this area.

III. EFFECTS ON MUSCLE

Muscle structure, function, and strength deteriorate with inactivity and immobilization, as demonstrated by findings from both experimental and clinical studies and during weightlessness. After short periods of immobilization, muscle disuse atrophy can be reversed fairly rapidly. But the longer the period of immobilization, the longer the time required to reverse the deterioration. With very prolonged immobilization (months or even years), there is little hope of totally reversing the problem, since viable muscle tissue is converted to a fibrous state. Exercise and electrostimulation of muscles have been studied as means of counteracting immobilization-induced muscle atrophy, but these techniques have not been entirely successful. More effective countermeasures must be sought that will be helpful to experimental subjects, clinically bed rested patients, the sedentary, the aging, and spaceflight participants on prolonged flights. Before such countermeasures can be developed, however, the mechanisms underlying muscular atrophy with immobilization must be more fully understood. Consequently, comprehensive research must still be conducted in this area, with a view toward answering the questions of underlying mechanisms and the development of reliable and effective countermeasures.

IV. METABOLIC AND ENDOCRINE CHANGES

Many metabolic and endocrine changes have been documented in individuals who become inactive because of disease, disability, aging, and exposure to weightlessness. To date, however, it has been impossible to assess the extent to which inactivity is responsible for these changes. Studies of healthy, bed rested individuals have shown early and rapid alterations in fluids and electrolytes and a reduction in blood volume. With continued inactivity, the change in blood volume is succeeded by an uncoupling of many neuroendocrine and neurohumoral regulatory mechanisms and by alterations in the number and affinity of receptors, those for insulin serving as a good example. The degree of such changes appears to be related to the physical condition of the individual prior to the onset of inactivity. Individuals who are highly conditioned by exercise training (particularly isotonic) show higher initial values in almost all categories and greatest absolute change.

Almost all 24-hr rhythms of the body show significant change (or shift) when subjects are confined to bed for significant periods. This is particularly the case for thyroid function or in the rhythms of most other endocrine variables, indicating that the activity level of the patient must be taken into account when single, fasting blood samples are drawn for diagnostic purposes. It cannot be assumed that observed changes are caused by disease or aging when they may very likely result from the environmental, physical, or emotional aspects of inactivity. Only a knowledge of the previous activity situation of the patient (e.g. a jogger, walker, or sedentary person) will allow the investigator to evaluate changes accurately.

Following prolonged bed rest, a loss of tolerance to stress is seen consistently. This loss may result from tissue hypoxia and increased pooling in the limbs. A decrease in oxygen content of the blood causes an increase in blood fibrinolytic activity, which has usually been attributed to the release of plasminogen activator in the vascular endothelium. It has been clearly documented that there is an increase in circulating fibrinogen with bed rest (Sandler and Winter, 1978). This finding is important for clinical as well as space considerations, because it is known that elevated fibrinogen levels create a greater tendency toward intravascular clotting. However, ambulatory controls confined to a restricted area also showed significantly elevated plasma fibrinogen and circulating fibrinogen after the study period. It would appear, therefore, that inactivity is as conducive to the observed changes as is immobilization.

The effects of prolonged inactivity or immobilization on the immune system have not been dealt with at length in this book, because reported findings have been inconclusive. Animal studies with rats suspended in the head-down position to reduce the effects of gravity have shown no significant impairment of the immune system (Caren et al., 1980). Studies with other species have shown that some animals of the same species were affected, while others were not. In one human study, Mikhaylovskiy and co-workers (1967) evaluated changes in the distribution of the immunoglobulins IgA, IgG, and IgM throughout the circulatory system with head-down bed rest. Six subjects bed rested for 5 days were catheterized for collecting blood samples from various locations in the cardiovascular system. Following bed rest, the investigators found less IgA in the venous blood from the brain than in arterial blood and more IgG in venous blood as compared with arterial blood. Distribution of IgM did not change. The observed changes were attributed to metabolic changes in the various organs.

Information from spaceflight experience has added little to our presently incomplete knowledge of the immunological effects of inactivity and weightlessness. However, no serious problems or significant changes in immunological factors have been reported. Too few humans have flown to allow conclusive judgments even where slight changes were seen. Sera from Apollo crewmen (Flights 7 through 13) were assayed for changes in serum proteins responsible for humoral immunity and immunoglobulins (Fischer *et al.*, 1972). Emphasis was placed on IgA, IgG, and IgM, as well as the third component of complement (C'3) and other serum proteins. After flight, IgG was decreased, whereas IgA and C'3 were elevated. The mild otitis seen in one crewman was attributed to the decrease in IgG. Astronauts on the 59–day Skylab Mission did not show the changes seen in the Apollo crewmen, although the samples were collected not only before and after flight, but also during flight. In fact, plasma proteins remained normal throughout flight and thereafter (Kimzey *et al.*, 1979). The question might be asked here as to whether the much greater mobility and exercise programs of the Skylab astronauts may have had some effect. Finally, crewmen of the Apollo–Soyuz Mission were treated for changes in microbial populations and humoral immunity, but they showed no demonstrable differences in specific antibody levels (DeCelle and Taylor, 1976).

V. PSYCHOSOCIAL ASPECTS OF IMMOBILIZATION

The restriction of mobility and the sensory deprivation occurring with prolonged inactivity or immobilization have been observed to result in mental disturbances (Wexler *et al.*, 1958). The monotony of prolonged bed rest can affect not only psychosocial responses but also central nervous system function (Ryback *et al.*, 1971a,b). For example, the theta rhythm of the hippocamus tends to disappear when there is a constant repetition of the same act or stimulus. Petukhov and Purakin (1968) have also reported a slowing of the electroencephalograph (EEG) brain waves in healthy subjects during prolonged bed rest. Since no single system of the body is entirely unaffected by changes in the other systems, it is reasonable to expect that documented changes in cardiovascular function, metabolism, muscle, and bone will also affect the neuropsychological condition of immobilized individuals.

The major early symptoms of deterioration with inactivity or immobilization are withdrawal, regression, lack of energy, decreased mental capacity, and exaggerated or inappropriate reactions to external conditions. These debilitating symptoms can be serious, but they can be offset to a certain degree by careful management of the immobilized individual. Important aids in offsetting psychosocial deterioration with immobilization include sensory stimulation, projects designed to improve motor and brain function, and exercise. These have all been used during long-term spaceflight. Soviet cosmonauts who were exposed to both weightlessness and confinement for very prolonged periods (up to 237 days) experienced a number of the symptoms of deterioration just outlined. However, the receipt of stimulating material from earth transported by unmanned space vehicles and the visits of manned crews for a week or so greatly improved their outlooks. They also reported that exercise gave them a feeling of well-being, even though it did not counteract their physical reentry problems. Since we have seen throughout this book that weightlessness provokes many of the same changes experienced with inactivity and immobilization on earth, further reports on humans who have spent a very long time in space should be helpful in coping with problems on earth.

Sensory stimulation is highly important in working with inactive or immobilized individuals. Television, radio, books, magazines, and newspapers should be provided, and the individual encouraged to use them. Time cues are also important for circadian rhythms and well-being of the individual who is inactive, immobilized, or confined, so clocks and calendars should always be provided.

Occupational therapy can improve motor and brain function and should be initiated. Getting the hands to work and the brain to function in accomplishing psychomotor tasks may also help to stimulate the intellectual, emotional, and social outlook of the individual.

Finally, physical inactivity cannot be allowed to continue very long, because a number of undesired effects can occur (see Chapters 3 and 4 on bone and muscle, respectively). Findings on the impact of exercise are variable (see Chapter 7). Nonetheless, a daily routine of exercise, where feasible, does provide a sense of well-being and can be useful in counteracting muscle atrophy and bone depletion. It also improves the individual's self image in that he or she has a feeling of some control over what is happening.

Environmental effects also may play an important role in alleviating the psychosocial effects of inactivity and immobilization. Full-spectrum lighting, which simulates the illumination of natural sunlight, is one environmental factor being considered. Hughes and Neer (1981) have reviewed the effects of lighting on the elderly whose problems are often those of inactivity or immobilization. The authors suggest that unsatisfactory illumination can contribute to fatigue, decreased performance, diminished immunological responses, and reduced physical fitness. Lighting affects the individual in two ways: (a) it provides information about the environment (visual), and (b) it affects photobiology through the skin directly or receptors situated there. Since lighting is a factor in the environment that can be modified fairly readily, further study in this area could provide information for offsetting the psychosocial deterioration that can occur with bed rest, clinically indicated immobilization, sedentary life styles and aging, as well as during spaceflight.

VI. THE IMPACT OF EXERCISE

Prolonged inactivity or immobilization results in a reduction in physical work capacity as manifested by decreases in both maximal and submaximal oxygen uptake. The amount of the decrease in maximal V_{O_2} depends on the length of the period of inactivity or immobilization and the aerobic fitness of the individual, regardless of age or sex. These problems following bed rest are usually associated with (a) reduced blood volume, submaximal and maximal stroke volume, maximal cardiac output, skeletal muscle tone and strength, and aerobic enzyme capacities and (b) increased venous compliance and submaximal and maximal heart rate. Normal function can be partially restored with regular muscular activity, orthostatic stress, or both, during the exposure to inactivity or immobilization. However, much more information is needed on the physical responses to exercise that occur under these conditions. A further understanding of the problem would benefit a broad spectrum of society: the sedentary, the aging, immobilized patients, and participants in long-term spaceflights.

With continued mechanization and technological aids a larger and larger number of people are being forced into or offered a more sedentary life style. An increasing number of our society is living longer and is manifesting the effects of the aging process. The human body adapts quite readily to a sedentary life style (as it does to the weightless environment). When this life continues with little or no exercise stress, physical work capacity decreases dramatically, and the decrease is associated with cardiovascular and metabolic changes similar to those accompanying exposure to bed rest. Again, the deterioration depends on the length of time the individual has been sedentary and on his or her physical fitness before the sedentary period began again, regardless of age. Although bed rest does not duplicate the sedentary condition, it has demonstrated physiological changes that are qualitatively similar. Thus, many of the findings from bed rest studies are applicable in this area. Results of these studies indicate that sedentary or aging individuals would do well to stand or walk periodically during the day to maintain a healthy physiological tolerance to exercise stress, even though this approach will not result in a higher physical work capacity.

Information on exercise tolerance following prolonged bed rest is also applicable to individuals immobilized for clinical reasons, especially for myocardial infarction. A particular problem exists in assessing the causes of physical deconditioning among immobilized patients: Does the deconditioning result from immobilization, or does it stem from disease factors? The preponderance of earlier bed rest studies were conducted using young, healthy individuals (primarily male). More recently, however, subjects in the older age groups (45–65 years, both male and female) and more prone to cardiovascular and other diseases have also been evaluated. Healthy subjects showed increased heart rate and heart-rate/arterial blood pressure during exercise following bed rest, although they manifested no deterioration in cardiac function. Findings for subjects with myocardial infarction (Convertino *et al.*, 1982; DeBusk *et al.*, 1983; Hung *et al.*, 1983) showed identical change and that orthostatic stress resulting from bed rest contributes to a decrease in cardiorespiratory capacity after myocardial infarction despite the extent of cardiac damage. It is important to remember, however, that heart rate and heart-rate/blood-pressure increases are important indicators of additional myocardial oxygen demands and, consequently, could forecast hazards for individuals suffering from chronic ischemic heart disease (Gobel *et al.*, 1978). Patients with significantly restricted myocardial blood flow also are at risk of angina pectoris and other clinical coronary events, and safe, reliable countermeasures should be developed for avoiding these risks. Current clinical practice for myocardial infarction patients emphasizes low-level exercise during the recovery process. However, recent studies of both bed rested subjects and clinical patients have indicated that the mere exposure to gravitational stress is equally beneficial (Convertino, 1983; DeBusk *et al.*, 1983; Hung *et al.*, 1983; Sivarajan *et al.*, 1981). Thus, available knowledge suggests that patients should sit up, stand, or walk regularly as soon as possible.

Most of what we know about the effects of inactivity or immobilization on exercise capacity has been obtained from studies of healthy, bed rested individuals and very physically fit space crews. These individuals have also been studied to determine whether exercise can be used as a countermeasure to offset the deconditioning inherent in altered gravity stress or its absence. As a countermeasure during these situations, neither isotonic nor isometric exercise has proved to be entirely effective. Further research is needed to determine the precise roots underlying exercise deconditioning with gravitational changes and to seek some means of offsetting this problem.

VII. SUMMARY

Although a vast amount of research has been conducted on inactivity and immobilization, as this book has attempted to indicate, there are many areas that still remain clouded. The bases of the observed changes that occur are still unclear in many cases, and the question of whether the changes may be permanent—given a prolonged period of gravity deprivation—still remains to be answered. Furthermore, we will need to know why some individuals are more severely affected than others. Answers to these questions will not be easily acquired. They will require many more hours of research with healthy bed rested individuals, clinically bed rested patients, the sedentary, the aging, and more research of flights of both humans and animals into the weightless environment.

REFERENCES

Caren, L.D., Mandel, A.D., and Nunes, J.A. (1980). *Aviat. Space Environ. Med.* **51**, 251-255.

Convertino, V.A. (1983). *J. Cardiac Rehab.* **3**, 660-663.

Convertino, V.A., Hung, J., Goldwater, D., and DeBusk, R.F. (1982). *Circulation* **65**, 134-140.

DeBusk, R.F., Convertino, V.A., and Goldwater, D. (1983). *Circulation* **68**, 245-250.

DeCelle, J.G. and Taylor, G.R. (1976). *Appl. Environ. Microbiol.* **32**, 659-665.

Fischer, E., Cress, R.H., Haines, G., *et al.* (1972). *Am. J. Phys. Med.* **50**, 230-234.

Gauer, O.H. and Henry, J.P. (1976). *In* "International Review Physiology, Cardiovascular Physiology" (A.C. Guyton and A.W. Cowley, eds.) Vol. 9, Sect. II, pp. 145-190. University Park Press, Baltimore, Maryland.

Gobel, F.L., Nordstrom, L.A., Nelson, R.R., Jorgensen, C.R., and Wang, Y. (1978). *Circulation* **57**, 549-556.

Hughes, P.C. and Neer, R.M. (1981). *Human Factors* **23(1)**, 65-85.

Hung, J., Goldwater, D., Convertino, V.A., McKillip, J.H., Goris, M.L., and DeBusk, R.F. (1983). *Am. J. Cardiol.* **51**, 344-348.

Kimzey, S.L., Leonard, J.T., and Johnson, P.C. (1979). *Acta Astronautica* **6**, 1289-1303.

Mikhaylovskiy, G.P., Benevolenskaya, T.V., Petrova, T.A., Yakoleva, I.Ya., Boykova, O.I., Kuz'min, M.P., Savilov, A.A., and Solov'yeva, S.N. (1967). *Space Biol. and Med.* **1(5)**, 86-90.

Petukhov, B.N. and Purakhin, Yu.N. (1968). *Space Biol. and Med.* **2**, 56-61.

Ryback, R.S., Lewis, O.F., and Lessard, C.S. (1971a). *Aerosp. Med.* **42**, 529-535.

Ryback, R.S., Trimble, R.W., Lewis, O.F., and Jennings, C.L. (1971b). *Aerosp. Med.* **42**, 408-415.

Sandler, H. and Winter, D.L. (1978). "Physiological Responses of Women to Simulated Weightlessness: A Review of the Significant Findings of the First Female Bed Rest Study." NASA SP-430, pp. 1-87.

Sivarajan, E.S., Bruce, R.A., Almes, M.J., Green, B., Belanger, L., Lindskog, B.D., Newton, K.M., and Mansfield, L.W. (1981). *N. Eng. J. Med.* **305**, 357-362.

Wexler, D., Mendelson, J., Leiderman, P.H., and Solomon, P. (1958). *AMA Arch. Neurol. and Psychiat.* **79**, 225-233.

Index